Unstable Modules over the Steenrod Algebra and Sullivan's Fixed Point Set Conjecture

Chicago Lectures in Mathematics Series
Robert J. Zimmer, Series Editor
J. Peter May, Spencer J. Bloch, Norman R. Lebovitz, William Fulton, and Carlos Kenig, editors

Other *Chicago Lectures in Mathematics* titles available from the University of Chicago Press:

Simplicial Objects in Algebraic Topology, by J. Peter May (1967)

Torsion-Free Modules, by Eben Matlis (1973)

Rings with Involution, by I. N. Herstein (1976)

Theory of Unitary Group Representation, by George V. Mackey (1976)

Infinite-Dimensional Optimization and Convexity, by Ivar Ekeland and Thomas Turnbull (1983)

Commutative Semigroup Rings, by Robert Gilmer (1984)

Navier-Stokes Equations, by Peter Constantin and Ciprian Foias (1988)

Essential Results of Functional Analysis, by Robert J. Zimmer (1990)

Fuchsian Groups, by Svetlana Katok (1992)

Topological Classification of Stratified Spaces, by Shmuel Weinberger (1994)

Lionel Schwartz

Unstable Modules over the Steenrod Algebra and Sullivan's Fixed Point Set Conjecture

The University of Chicago Press
Chicago and London

LIONEL SCHWARTZ is professor of mathematics at the
Institut Galilée, Université Paris-Nord.

The University of Chicago Press, Chicago 60637
The University of Chicago Press, Ltd., London

© 1994 by The University of Chicago
All rights reserved. Published 1994
Printed in the United States of America

03 02 01 00 99 98 97 96 95 94   1 2 3 4 5

ISBN: 0-226-74202-4 (cloth)
         0-226-74203-2 (paper)

Library of Congress Cataloging-in-Publication Data

Schwartz, Lionel.
    Unstable modules over the Steenrod algebra and Sullivan's fixed
point set conjecture / Lionel Schwartz.
         p.   cm. — (Chicago lectures in mathematics series)
    Includes bibliographical references (p.   –   ) and indexes.
    1. Steenrod algebra.   2. Modules (Algebra)   3. Fixed point theory.
I. Title.   II. Series: Chicago lectures in mathematics.
QA612.782.S39   1994
512'.55—dc20   93-48596

♾ The paper used in this publication meets the
minimum requirements of the American National Standard
for Information Sciences—Permanence of Paper for
Printed Library Materials, ANSI Z39.48-1984.

This volume was composed in Times and MathTime fonts using
Lams TEX, osudeG.sty and Sweet-teX at Univ. de Paris-Sud.
The master copy was flashed on a Linotronic 1270 dpi typesetter at
Aston University.

à mes parents

Contents

Part 1

The algebraic structure of the category $\mathcal{U}$ and the functor T_V

Part 2

Deeper algebraic structure

Part 3

The Sullivan conjecture and the cohomology of mapping spaces

Frequently used notation

$\mathcal{A}$ — is the mod p Steenrod algebra

$\mathcal{A}_*$ — is the dual of the mod p Steenrod algebra

$\mathcal{A}lg$ — is the category of unital, commutative, $\mathbb{N}$-graded $\mathbb{F}_p$-algebras

$\mathcal{A}lg_a$ — is the category of augmented objects in $\mathcal{A}lg$

$\mathcal{E}$ — is the category of $\mathbb{F}_p$-vector spaces

$\mathcal{E}_{\mathrm{gr}}$ — is the category of graded $\mathbb{F}_p$-vector spaces

H^*X — is the mod p cohomology of the space X

$\widetilde{H}^*X$ — is the reduced mod p cohomology of the space X

$H^*V \cong H^*BV$ — is the mod p cohomology of V

$\widetilde{H}^*V$ — is the reduced mod p cohomology of V

$\mathcal{K}$ — is the category of unstable $\mathcal{A}$-algebras

$\mathcal{K}_a$ — is the category of augmented objects in $\mathcal{K}$

$M^{\oplus a}$, $a \in \mathbb{N}$ (or more generally a cardinal) — denotes the direct sum of a copies of the object M of $\mathcal{E}$, or of $\mathcal{E}_{\mathrm{gr}}$, or of $\mathcal{U}$ etc.

$\mathcal{O}$ — denotes always "a" forgetful functor

$\mathcal{U}$ — is the category of unstable $\mathcal{A}$-modules

$\mathcal{U}'$ — is the category of evenly graded unstable $\mathcal{A}$-modules

V, W, ... — denote finite dimensional $\mathbb{F}_p$-vector spaces

$\alpha(n)$, $n \in \mathbb{N}$ — is the sum of the coefficients in the p-adic expansion of the integer n

$\twoheadrightarrow$ — always denotes an epimorphism

$\hookrightarrow$ — always denotes a monomorphism

Introduction

Since their construction by Steenrod, the operations Sq^i and P^i have played a central role in homotopy theory. The Steenrod algebra is certainly one of the most efficient tools in the hands of algebraic topologists. The following notes are primarily based on a course given at the University of Chicago (in the Winter term of 1988) and on a course given at Northwestern University (in the Spring term of 1988). They are concerned with the theory of unstable modules over the mod p Steenrod algebra $\mathcal{A}$, its recent developments and its spectacular applications to the homotopy and homology theory of spaces of mappings of certain classifying spaces. These developments found their origin and their motivation in a conjecture of D. Sullivan, and in a conjecture of G. Segal.

D. Sullivan's fixed point set conjecture is concerned with the analysis of the fixed point set of a group action. The question is to decide to what extent the homotopy type of the fixed point set of a finite group acting 'nicely' on a finite complex (for example the action by complex conjugation of the group $\mathbb{Z}/2$ on a complex algebraic variety X) is determined by a 'homotopy model'.

Let us be more precise. Let X be a *finite* $\mathbb{Z}/2$-CW-complex. One defines the homotopy fixed point $X^{h\mathbb{Z}/2}$ to be the set of $\mathbb{Z}/2$-equivariant maps $\mathrm{map}_{\mathbb{Z}/2}(S^\infty, X)$, the sphere S^∞ being provided with the antipodal action. Such an equivariant map is called a homotopy fixed point. An honest fixed point obviously determines a homotopy fixed point and there is a map from the fixed point set $X^{\mathbb{Z}/2}$ to the homotopy fixed point set

1

$$X^{\mathbb{Z}/2} \to X^{h\mathbb{Z}/2}.$$

D. Sullivan conjectured that this is a homotopy equivalence after suitable completions on both sides [**Su**]. Even the particular case when the action of the group is trivial is far from being obvious. Indeed it reduces to showing that the space of pointed maps, $\mathrm{map}_*(B\mathbb{Z}/2, X)$, is contractible. This was considered as a test case by Sullivan.

G. Segal's Burnside ring conjecture is related to the stable cohomotopy of the classifying space BG of a finite group G. It is an analog of the Atiyah-Segal completion theorem describing the K-theory of BG as the completed representation ring of G. A weak form of the conjecture describes $\pi_S^0(BG)$ as the Burnside ring of G completed with respect to the topology defined by the augmentation ideal. Recall that the Burnside ring is the Grothendieck ring of finite G-sets under disjoint union and cartesian product.

These notes are concerned, for the homotopy theory part with Sullivan's conjecture and related topics.

In the first part of the book, i.e. Chapter 1 to 3, we develop the basic algebraic structure of the category of unstable modules over the Steenrod algebra.

Recall that a module M over the Steenrod algebra is said to be unstable if:

— $\mathrm{Sq}^i x = 0$ for all $x \in M$ such that $|x| < i$ if $p = 2$;

— $\beta^e P^i x = 0$ for all $x \in M$ such that $|x| < e + 2i$, $e = 0, 1$, if $p > 2$.

Here $|x|$ denotes the degree of x.

The (abelian) category of unstable $\mathcal{A}$-modules is denoted by $\mathcal{U}$.

The mod p cohomology, H^*X, of a space X is an unstable $\mathcal{A}$-module. Moreover it is an unstable $\mathcal{A}$-algebra. It has an algebra product $H^*X \otimes H^*X \to H^*X$ which is an $\mathcal{A}$-linear map, and moreover $Sq^{|x|}x = x^2$ for any class x if $p = 2$, $P^{|x|/2}x = x^p$ for any class x of even degree if $p > 2$.

The category of unstable $\mathcal{A}$-algebras is denoted by $\mathcal{K}$.

In Chapter 1, we recollect basic and classical facts about the categories $\mathcal{U}$ and $\mathcal{K}$.

In Chapter 2, we describe the basic injective objects of the category $\mathcal{U}$. There are three basic types of injective objects.

The first ones are the canonical cogenerators denoted by $J(n)$, $n \geq 0$; these are objects representing the functor from $\mathcal{U}$ to the category $\mathcal{E}$ of $\mathbb{F}_p$-vector spaces given by $M \mapsto \mathrm{Hom}(M^n, \mathbb{F}_p)$. The unstable $\mathcal{A}$-modules $J(n)$ are injective in $\mathcal{U}$ because the functor $M \mapsto M^{n*}$ is left exact. We note that the unstable $\mathcal{A}$-modules $J(n)$ were known for a long time in a slightly different form. They are isomorphic to the mod p cohomology of the n-th Spanier-Whitehead dual of the n-th Brown-Gitler spectrum [**BG**][**BC**][**Mh**]. We shall call them Brown-Gitler modules. They are described in Sections 2.1 to 2.4, relations with Milnor's dual of the Steenrod algebra are discussed in Section 2.5.

In his work on the Segal conjecture for elementary abelian 2-groups [**Cl1**] Carlsson considered certain unstable $\mathcal{A}$-modules that are filtered inverse limits of the unstable $\mathcal{A}$-modules $J(n)$. These modules, denoted by $K(i)$ are 'obvious cogenerators' for 'reduced' unstable $\mathcal{A}$-modules. An unstable $\mathcal{A}$-module is said to be reduced if it does not contain a non-trivial suspension; at $p = 2$ it means that the map $x \mapsto Sq^{|x|}x$ is injective. For an unstable $\mathcal{A}$-algebra K, again at the prime 2, it means that K does not contain nilpotent elements. Formal arguments show that the unstable $\mathcal{A}$-modules $K(i)$ are injective. We shall call the $K(i)$'s Carlsson modules. This is discussed in Sections 2.6 and 2.7.

The last type is given in Section 2.8, it is the one considered by J. Lannes and S. Zarati [**LZ1**]. The tensor product of Brown-Gitler modules by Carlsson modules. In Section 2.9 we discuss relations of the Carlsson's module with the combinatoric of binary trees.

In Chapter 3, we prove the injectivity of the mod p-cohomology of elementary abelian p-groups and we introduce Lannes' functor T_V. Then we discuss the most important properties of T_V.

Recall that Carlsson's key observation in his work on the Segal conjecture for elementary abelian 2-groups is as follows. The mod 2 cohomology of the group $\mathbb{Z}/2$, $\tilde{H}^*B\mathbb{Z}/2 \cong \tilde{H}^*\mathbb{Z}/2 \cong u\mathbb{F}_2[u]$, $|u| = 1$, is a direct summand of the unstable $\mathcal{A}$-module $K(1)$ alluded to above, more generally $\tilde{H}^*B(\mathbb{Z}/2^{\oplus n}) \cong \tilde{H}^*(\mathbb{Z}/2^{\oplus n})$ is a direct summand of $K(1)^{\otimes n}$. H. Miller observed that this means that

$H^*\mathbb{Z}/2$ is injective in the category $\mathcal{U}$ and extended the analysis to all primes p. Then Lannes and Zarati extended the Carlsson-Miller analysis to the unstable $\mathcal{A}$-modules $H^*V \otimes J(n)$, for all V, and all $n \geq 0$. All this is covered in Section 3.1.

In Section 3.2 we describe the construction of Lannes' functor T_V. Freyd's adjoint functor theorem implies that for any unstable $\mathcal{A}$-module L which is of finite dimension in each degree (i.e. which is of finite type) the functor $\mathcal{U} \rightarrow \mathcal{U}$, $M \mapsto L \otimes M$ has a left adjoint $M \mapsto (M : L)_{\mathcal{U}}$ that deserves to be called a 'division functor'. There is an analogous 'division functor' $K \mapsto (K : L)_{\mathcal{K}}$ in the category of unstable $\mathcal{A}$-algebras. Lannes considered the left adjoint of the functor $M \mapsto H^*V \otimes M$ [**L1**][**L4**]. He observed that the injectivity of $H^*V \otimes J(n)$ translates immediately into the exactness of the functor $M \mapsto (M : H^*V)_{\mathcal{U}}$. He denoted this functor by T_V. In Sections 3.3 and 3.4 we give some explicit computations with T_V.

In Sections 3.5 to 3.7, we consider the second main property of the functor T_V. There is a natural map

$$T_V(M \otimes N) \rightarrow T_V(M) \otimes T_V(N),$$

induced by the algebra product $H^*V \otimes H^*V \rightarrow H^*V$. This natural map is an isomorphism for all M and N in the category $\mathcal{U}$. We show in Section 3.8 that this implies T_V induces a functor from $\mathcal{K}$ to itself which is equivalent to the division functor $(: H^*V)_{\mathcal{K}}$ (defined as above).

The preceding 'tensor product theorem' is related to the Adams-Gunawardena-Miller work on the Segal conjecture [**AGM**] and to another work of Lannes [**LZ2**]. The work of Adams Gunawardena and Miller was concerned with the computation of the groups $\text{Ext}_{\mathcal{A}}^s(H^*V, H^*W \otimes M)$, where V and W are finite dimensional $\mathbb{F}_p$-vector spaces and M is an $\mathcal{A}$-module which is bounded below but not necessarily unstable. It turns out that the case $s = 0$ of the computation itself is far from being obvious and is of special interest.

In this case the result states that the linear extension of the map

$$f \mapsto (Bf)^*,$$

$$\text{Hom}(W, V) \rightarrow \text{Hom}_{\mathcal{A}}(H^*V, H^*W) = \text{Hom}_{\mathcal{U}}(H^*V, H^*W)$$

is a linear isomorphism. In particular, if $V = W$, there is an isomorphism of rings

$$\mathrm{End}_A(H^*V) \cong \mathrm{End}_{\mathcal{U}}(H^*V) \cong \mathbb{F}_p[\mathrm{End}\, V]^{\mathrm{opp}}$$

where $\mathbb{F}_p[\mathrm{End}\, V]^{\mathrm{opp}}$ is the opposite of the monoid ring $\mathbb{F}_p[\mathrm{End}\, V]$.

These last results were proved around the same time by Lannes. He proved a result that looks a bit more general, but that is equivalent to the preceding one [LZ2]. Let K be an unstable $\mathcal{A}$-algebra. The set $\mathrm{Hom}_{\mathcal{K}}(K, H^*V)$ injects into the $\mathbb{F}_p$-vector space $\mathrm{Hom}_{\mathcal{U}}(K, H^*V)$. Then the (unique) linear extension of this injection, from the $\mathbb{F}_p$-vector space freely generated by $\mathrm{Hom}_{\mathcal{K}}(K, H^*V)$:

$$\mathbb{F}_p\left\{\mathrm{Hom}_{\mathcal{K}}(K, H^*V)\right\} \rightarrow \mathrm{Hom}_{\mathcal{U}}(K, H^*V)$$

is an isomorphism (this is true as stated if $\mathrm{Hom}_{\mathcal{K}}(K, H^*V)$ is finite; otherwise one has to take into account profinite topologies). This has to be thought of as a 'linearization principle' and this implies the injectivity of H^*V (in the categorical sense) in $\mathcal{K}$. We note that special cases of this result were present in the work of J.F. Adams and C. Wilkerson [AW].

In Sections 3.9 and 3.10, we give further examples.

In Sections 3.11 to 3.13, we show that any indecomposable injective object of $\mathcal{U}$ occurs as a direct summand of some $H^*V \otimes J(n)$. In particular any indecomposable reduced injective object of $\mathcal{U}$ occurs as a direct summand of some H^*V. These results allow us to characterize unstable modules M for which $T_V M$ is k-connected. Moreover this implies that the unstable $\mathcal{A}$-module module $K(i)$ can be written uniquely as a direct sum of indecomposable summands of H^*V's. This was proved by Lannes and the author in [LS2]. H. Campbell and P. Selick proved later another result relating the unstable $\mathcal{A}$-modules $K(i)$ more precisely to H^*V [CS]. Their proof uses the 'extended mod p' Steenrod algebra.

The second part of the book is concerned with later algebraic developments.

In Chapter 4, we describe some results about the structure of the indecomposable summands of H^*V. We show, following D. Carlisle and N. Kuhn [CK], how to classify them by using modular representation theory (Section 4.1). Information about their Poincaré series and their $\mathcal{A}$-module structure is given. The decomposition of the unstable

$\mathcal{A}$-modules $K(i)$ in terms of indecomposable summands of H^*V is given (Section 4.3).

In Chapter 5, we consider the quotient category $\mathcal{U}/\mathcal{N}il$. The full subcategory $\mathcal{N}il$ of $\mathcal{U}$ is the smallest Serre class that contains suspensions and is stable under colimit. We prove that $\mathcal{U}/\mathcal{N}il$ is equivalent to the category $\mathcal{F}_\omega$ of 'analytic functors' from the category $\mathcal{E}_f$ of finite dimensional $\mathbb{F}_p$-vector spaces to $\mathcal{E}$ **[EM][Md]**. The equivalence $f : \mathcal{U}/\mathcal{N}il \to \mathcal{F}_\omega$ is induced by the functor

$$\mathcal{U} \to \mathcal{F} = \mathcal{E}^{\mathcal{E}_f} , \quad M \mapsto \left(V \mapsto (T_V M)^0 \right) .$$

The difficulty is to show that all analytic functors are 'in the image' of f. The problem is that the obvious cogenerators for $\mathcal{U}/\mathcal{N}il$ are the unstable $\mathcal{A}$-modules $K(i)$, and that they have no easy link with the obvious cogenerators for $\mathcal{F}$ which are the functors $V \mapsto \mathbb{F}_p^{\mathrm{Hom}\,(V,W)}$ (W is fixed) that represent $F \mapsto F(W)^*$, $\mathcal{F} \to \mathcal{E}$. The proof presented in these notes (Section 5.1 to 5.4) depends on internal properties of $\mathcal{F}_\omega$ and not (as the original one **[HLS2]**) on properties of f with respect to certain products.

In Section 5.5, we introduce a certain filtration on $\mathcal{U}/\mathcal{N}il$ that identifies with the degree filtration on $\mathcal{F}_\omega$. The quotient category $\mathcal{V}_n/\mathcal{V}_{n-1}$ is shown to be equivalent to the category of modules over the group ring $\mathbb{F}_p[\mathfrak{S}_n]$ of the symmetric group $\mathfrak{S}_n$ **[LS4][FS]**. As an application we give the decomposition of $K(i)$ into indecomposable direct summands (Section 5.7).

We note also that there is a decreasing filtration on $\mathcal{U}/\mathcal{N}il$ and $\mathcal{F}_\omega$ by subcategories $\mathcal{L}_i$ such that $\bigcap_d \mathcal{L}_d = \{0\}$ and such that the quotient category $\mathcal{L}_d/\mathcal{L}_{d+1}$ is equivalent to the category of modules over the group ring $\mathbb{F}_p[GL_d\mathbb{F}_p]$. This is related to previous work of J.C. Harris and N. Kuhn **[HK]**.

In Chapter 6, we describe a filtration on the category $\mathcal{U}$ **[Sc1]**. Let $\mathcal{N}il_i$ $i \geq 0$ be the smallest 'Serre class' in $\mathcal{U}$ that contains all i-suspensions and is stable under colimits ($\mathcal{N}il_0 = \mathcal{U}$, $\mathcal{N}il_1$ has been denoted by $\mathcal{N}il$). We show in Section 6.1 that there are equivalences of quotient categories:

$$\mathcal{N}il_i/\mathcal{N}il_{i+1} \cong \mathcal{N}il_{i-1}/\mathcal{N}il_i , \quad i \geq 1 .$$

As a consequence $\mathcal{U}/\mathcal{N}il$ should be thought of as a 'building block' for $\mathcal{U}$. Its study is closely related to that of reduced unstable $\mathcal{A}$-modules. We consider closely related subcategories, denoted by $\bar{\mathcal{N}il}_\ell$ of $\mathcal{U}$ and describe the Krull filtration on $\mathcal{U}$ (Section 6.2). Then we consider localization away from the category $\mathcal{N}il_i$ (Section 6.3) and give some applications.

In part 3, we consider homotopical applications of the algebraic results of the first part. We start by describing the content of Chapters 8 and 9.

The first homotopical application has been in fact the motivation of the later developments. It is H. Miller's celebrated theorem on the contractibility of the space of pointed maps $\mathrm{map}_*(B\mathbb{Z}/p, X)$, where X is a finite dimensional complex. Taking into account the injectivity of $H^*\mathbb{Z}/p$ in the category $\mathcal{U}$, Miller was able to prove this particular case of the Sullivan conjecture for spaces with 'very nice' cohomology by using Massey-Peterson towers [Ml1] . Then he was able to prove the general case using the cosimplicial technology of A.K. Bousfield and D. Kan. These authors have constructed a homotopy spectral sequence for the homotopy of the 'total space' of a pointed cosimplicial space [BK2] that, as supplemented by E. Dror, W. Dwyer and D. Kan, works very well in the present context. To show that the space $\mathrm{map}_*(B\mathbb{Z}/p, X)$ is contractible (X finite, nilpotent) it is enough to prove the two following facts.

(i) The set $\mathrm{Hom}_{\mathcal{K}}(H^*X, H^*\mathbb{Z}/p)$ is the one point set.

(ii) The (comonad-)derived functors

$$\mathrm{Ext}^s_{\mathcal{K}}(H^*X, \Sigma^t(\tilde{H}^*\mathbb{Z}/p)^+) \ , \quad t \geq s \geq 0 \ , \quad t > 0$$

of the functor $\mathcal{K} \to \mathcal{E}$, $K \mapsto \mathrm{Hom}_{\mathcal{K}}(K, \Sigma^t(\tilde{H}^*\mathbb{Z}/p)^+)$ are trivial.

The expression $\Sigma^t(H^*\mathbb{Z}/p)^+$ denotes the unstable $\mathcal{A}$-algebra which is equal to $H^n\mathbb{Z}/p$ in degree $n+t$, $n > 0$, to $\mathbb{F}_p$ in degree zero, to $\{0\}$ in degrees 1 to t and which is provided with the trivial product.

The first fact is trivial to check.

The second fact is much harder to check. Thom's work on unoriented cobordism — for the algebraic part — was essentially the computation of the E_2-term of a certain Adams spectral sequence (although the spectral sequence was not defined at that time!). The spectral sequence in question collapses at E_2 because the E_2-term, $\mathrm{Ext}_{\mathcal{A}}^s (H^* MO, \Sigma^t \mathbb{F}_2)$, is trivial in cohomological dimension ≥ 1. This is because the mod 2 cohomology of the Thom spectrum MO is $\mathcal{A}$-free. In the context of the Sullivan conjecture an analogous situation occurs: the right entry in the Ext-group under consideration is closely related to an injective object in $\mathcal{U}$ and the category $\mathcal{K}$ is closely related to the category $\mathcal{U}$. Miller showed how to reduce the computation of these Ext-groups in the category $\mathcal{K}$ to a computation of Ext-groups in the category $\mathcal{U}$. Then using both the injectivity of $H^*\mathbb{Z}/2$ and the boundness hypothesis on H^*X he was able to show the triviality of the latter groups.

We describe a proof of this theorem in Chapter 8 (Section 8.6). In fact we show that the mapping space $\mathrm{map}_*(B\mathbb{Z}/p, X)$ (X is no longer finite) is k-connected if and only if the cohomology H^*X is in $\overline{Nil}_{k+1}$. This, under reasonable assumptions on X. In particular this leads to a converse of Miller's theorem: if $\mathrm{map}_*(B\mathbb{Z}/p, X)$ is contractible (H^*X of finite type but not necessarily bounded, and $\pi_1 X = \{0\}$), then H^*X is locally finite as an $\mathcal{A}$-module, i.e. the span of any element over the Steenrod algebra is finite.

The second homotopical application that we shall give is a conjecture of Miller. We describe it in Sections 8.1 to 8.4. The conjecture states as follows.

Let X be a space with 'mild properties'. Then the map

$$[BV, X] \to \mathrm{Hom}_{\mathcal{K}}(H^*X, H^*BV) , \quad f \mapsto f^*$$

is bijective. Miller's theorem is evidence for that. Using Massey-Peterson towers, Miller proved this conjecture if X has 'very nice' cohomology [Ml1]. Later on Lannes and Zarati proved it for X an infinite loop space. Finally Lannes proved the general case.

This result is proved by using the Bousfield-Kan technology, as generalized by Bousfield [Bo4]. The 'classical' Bousfield-Kan technology deals with the component of the trivial map of a mapping space. In a letter to J. Neisendorfer (1985), later developed in [Bo4], Bousfield showed how to solve the following two problems:

(i) Given spaces X and Y and $\varphi \in \mathrm{Hom}_{\mathcal{K}}(H^*Y, H^*X)$, when does there exist a map $f : X \to Y$ inducing φ in mod p cohomology?

(ii) Are two such maps homotopic?

The idea is to consider the total space of the cosimplicial space of maps, from X to a resolution of Y, that, in some adequate sense, 'induce φ in mod p cohomology'. We give the necessary information in Sections 8.2 and 8.3. The conjecture of Miller is proved in Section 8.4.

In Section 8.7, we give, using the Eilenberg-Moore spectral sequence, applications of T_V to the cohomology of fibrations [Sc1] [LS4]. For example, for a 'nice' fibration

$$F \to E \to B \,,$$

one gets that the kernel of the edge homomorphism of the spectral sequence

$$H^*E \underset{H^*B}{\otimes} \mathbb{F}_2 \to H^*F$$

consists of nilpotent elements. This leads to a generalization of a theorem of Serre [LS4].

In Chapter 9, we prove the general form of the Sullivan conjecture following the approach of Lannes. Let X be a finite $\mathbb{Z}/p$ -CW-complex. Sullivan's conjecture as proved by Carlsson, Lannes and Miller (the hypotheses are a bit different in Carlsson's work) states as follows. Let $X \mapsto \mathbb{F}_{p\infty} X$ denote the Bousfield-Kan p-completion functor. Let X be a finite $\mathbb{Z}/p$ -CW-complex. There is a natural map

$$\mathbb{F}_{p\infty}(X^{\mathbb{Z}/p}) \to (\mathbb{F}_{p\infty} X)^{h\mathbb{Z}/p}$$

which is a homotopy equivalence. The result holds when one replaces $\mathbb{Z}/p$ by any finite p-group π.

We give Lannes' proof (Section 9.1 to 9.4). It depends on the analysis of the mapping space from $B\mathbb{Z}/p$ to the Borel construction on X:

$$\mathrm{map}(B\mathbb{Z}/p, E\mathbb{Z}/p \underset{\mathbb{Z}/p}{\times} X) \,.$$

Let

$$\mathrm{map}_1(B\mathbb{Z}/p, E\mathbb{Z}/p \underset{\mathbb{Z}/p}{\times} X)$$

be the subspace consisting of maps φ such that the composite

$$B\mathbb{Z}/p \xrightarrow{\varphi} E\mathbb{Z}/p \underset{\mathbb{Z}/p}{\times} X \xrightarrow{\pi} B\mathbb{Z}/p$$

induces the identity on the fundamental group. There is an obvious inclusion

$$B\mathbb{Z}/p \times X^{\mathbb{Z}/p} \subset \operatorname{map}_1(B\mathbb{Z}/p, E\mathbb{Z}/p \underset{\mathbb{Z}/p}{\times} X).$$

Using a Bousfield-Kan' type comparison theorem for appropriate homotopy spectral sequences on both sides, one shows that this inclusion becomes an equivalence after suitable completions. The comparison of the E_2-terms of the spectral sequences depends on the adjointness properties of T_V.

All that precedes focuses attention on the problem of computing $H^* \operatorname{map}(BV, X)$ and suggests strongly that it should be isomorphic to $T_V H^* X$, as conjectured by Lannes. We prove this 'mapping space theorem' in Sections 9.7 to 9.11. Lannes proved it by using Bousfield's work [Bo3] on the homology spectral sequence of a cosimplicial space combined with T_V techniques. Again the spectral sequence under consideration collapses at E_2. Aside from the usual hypotheses on X (nilpotence, finite fundamental group) one needs to know that $T_V H^* X$ is finite dimensional in all degrees and trivial in degree 1. Lannes analysis is delicate and we shall not reproduce it here. We describe the beautiful approach of E. Dror-Farjoun and J. Smith [DS] (Sections 9.7 to 9.10). Their idea is as follows. The result is easy if the target space is an Eilenberg-MacLane space. Therefore one would like to try an induction on the Postnikov tower. One proceeds as follows. Let $F \to E \to B$ be a 'nice' fibration (for example F, E and B are of finite type, and $\pi_1 B = \{0\}$). One compares the Eilenberg-Moore spectral sequence of the fibration

$$\operatorname{map}(BV, F) \to \operatorname{map}(BV, E) \to \operatorname{map}(BV, B)$$

of mapping spaces, and the spectral sequence obtained by applying the functor T_V to the Eilenberg-Moore spectral sequence of the original fibration. There is a map from the last one to the first one that happens to be an isomorphism on the $E_2^{s,t}$-term as soon as $H^* E^{BV} \cong T_V H^* E$ and $H^* B^{BV} \cong T_V H^* B$. It follows from the classical comparison theorem that one also has an isomorphism $H^* F^{BV} \cong T_V H^* F$.

It should be stressed that this approach gives also a new proof of Lannes' theorem on the classification, up to homotopy, of maps from BV to X.

We come back to Chapter 7. This chapter gives the information about derivations and their derived functors, i.e. about André-Quillen cohomology, which is necessary to the analysis of the E_2 terms of the spectral sequences that we consider. We need a certain cancellation theorem that ensures the triviality of (most) of the E_2 terms under consideration.

Bousfield's condition for the existence of a map $X \to Y$ representing a given morphism $H^*Y \to H^*X$ is the triviality of the ' -1 ' column of the homotopy spectral sequence of the cosimplicial space. This $E_2^{s,s-1}$-term defined for $s \geq 2$ is a certain 'comonad'-derived functor

$$H^*X \mapsto \mathrm{Der}_{\mathcal{K}}(H^*X, \Sigma^{s-1}H^*V)$$

of the functor

$$H^*X \mapsto \mathrm{Der}_{\mathcal{K}/H^*V}(H^*X, \Sigma^{s-1}H^*V)$$

which is the $\mathbb{F}_p$-vector space of 'derivations' of degree $-s + 1$ from H^*X to H^*V with respect to the given map $\varphi : H^*X \to H^*V$ (Section 7.7). If φ is the trivial algebra map it reduces to the usual functor $\mathrm{Hom}_{\mathcal{K}}(H^*X, \Sigma^{s-1}\tilde{H}^*V^+)$.

We show how Lannes' functor T_V fits in this context. By adjointness and the fundamental properties of T_V, one shows that

$$\mathrm{Der}_{\mathcal{K}/H^*V}^s(H^*X, \Sigma^t H^*V)$$

is isomorphic to

$$\mathrm{Ext}_{\mathcal{K}_a}^s(T_V H^*X, \Sigma^t (\mathbb{F}_p)^+),$$

where $\mathcal{K}_a$ denotes the category of augmented objects of $\mathcal{K}$. This is proved in Section 7.8. The triviality result one is looking for reduces to a connectivity theorem for these groups which is, if not classical, at least expected. Indeed one has to show that for any augmented unstable $\mathcal{A}$-algebra K the groups

$$\mathrm{Ext}_{\mathcal{K}_a}^s(K, \Sigma^t (\mathbb{F}_p)^+)$$

are trivial for $t = s - 1$, $s \geq 2$ and $t = s$, $s \geq 1$.

The proof of this connectivity theorem is given in Sections 7.3 and 7.4. In Section 7.6 we give a change of rings theorem for these Ext-groups.

We note that these groups of derivations as well as the E_2-term of the Bousfield-Kan spectral sequence have subsequently been intensively studied by P. Goerss [Ge1][Ge2].

We conclude by surveying later developments that are not included in the book.

The mapping space theorem has important applications. Among them is the description by Lannes [L2] of the space map(BV, BG) for G a compact Lie group. This was also obtained by W. Dwyer and A. Zabrodsky [DZ] using the Sullivan conjecture. The statement is given (not proved) in Section 9.6. The mapping space theorem and (or) the Sullivan conjecture are essential ingredients in the computation of self-maps of the classifying space, BG, of a simple compact Lie goup G by S. Jackowski, J. McClure and R. Oliver [JMO]. Recall also that the computation of $[BV, BG]$ was used by K. Ishiguro [I] in his proof of Sullivan's conjecture concerning the non-existence of certain unstable Adams operations. The mapping space theorem has also been of particular importance in the work of various people for constructing old and new finite H-spaces (Aguade, Lannes, Dwyer-Wilkerson [DW3]). This has also been used in the problem of homotopy uniqueness and homotopy decomposition of classifying spaces (Dwyer, Miller, C. Wilkerson, [DMW][DW1][DW4]).

The work on analytic functors of Chapter 5 was extended in [HLS2] to the category $\mathcal{K}/\mathcal{N}il$ (unstable $\mathcal{A}$-algebras modulo nilpotent unstable $\mathcal{A}$-algebras). This leads to new proofs of the Adams-Wilkerson theorems [AW] as well as generalizations of these theorems. We note also that in [HLS1] various applications of these results to group cohomology are given.

Finally, we mention some other results. In [H1] Henn has characterized all finite groups whose mod p cohomology is injective. D. Benson and V. Franjou gave a very simple and elegant proof of the injectivity of $H^*\mathbb{Z}/p$ in $\mathcal{U}$ that depends on the lattice structure of subobjects of $H^*\mathbb{Z}/p$ in $\mathcal{U}/\mathcal{N}il$ [BF]. Also N. Kuhn [K1] has shown that the fact that the functors $H_n : V \mapsto \left(V^{\otimes n}\right)^{\mathfrak{S}_n}$ are generators for the category $\mathcal{F}_\omega$ together with a general form of Morita equivalence yield a new proof of the injectivity of H^*V as well as a new proof of the Adams-Gunawardena-Miller, Lannes theorem.

In conclusion, I would like to express my thanks to J.-P. May who proposed I publish these lectures as a volume of the University of Chicago 'blue' series.

I would like to thank J. Lannes and H. Miller for so many discussions about their work.

Among all those who helped me during the preparation of these notes, special thanks are due to J. Harris, N. Kuhn, and M. Zisman for their useful advice and comments.

It is a pleasure to acknowledge the support of the URA 1169 of the CNRS, and of its director L. Siebenmann who guided the preparation of the master copy. In particular, I would like to thank Mme B. Barbichon for her careful and beautiful TEX work. Thanks are also due to Aston University, where the master was flashed under the auspices of the UK TEX archive, and finally to the University of Chicago Press for constant friendly encouragement.

PART
1
ONE

The algebraic structure of the category $\mathcal{U}$

and the functor T_V

1. Recollections concerning the Steenrod algebra and unstable $\mathcal{A}$-modules

1.1. The Steenrod algebra

Let p be a prime number. The mod p Steenrod algebra $\mathcal{A}$ is the quotient of the 'free' associative unital graded $\mathbb{F}_p$-algebra generated by the elements:

Sq^i of degree i, $i > 0$, if $p = 2$,

β of degree 1 subject to $\beta^2 = 0$ and

P^i of degree $2i(p-1)$, $i > 0$, if $p > 2$;

by the ideal generated by the elements, known as the **Adem relations**

$$\mathrm{Sq}^i\,\mathrm{Sq}^j - \sum_0^{[i/2]} \binom{j-k-1}{i-2k} \mathrm{Sq}^{i+j-k}\,\mathrm{Sq}^k$$

for all $i, j > 0$ such that $i < 2j$ if $p = 2$;

$$P^i P^j - \sum_0^{[i/p]} (-1)^{i+t} \binom{(p-1)(j-t)-1}{i-pt} P^{i+j-t} P^t$$

for all $i, j > 0$ such that $i < pj$ and

$$P^i \beta P^j - \sum_0^{[i/p]} (-1)^{i+t} \binom{(p-1)(j-t)}{i-pt} \beta P^{i+j-t} P^t$$

$$- \sum_0^{[(i-1)/p]} (-1)^{i+t-1} \binom{(p-1)(j-t)-1}{i-pt-1} P^{i+j-t} \beta P^t$$

for all $i, j > 0$ such that $i \leq pj$ if $p > 2$.

In these formulas Sq^0 (resp. P^0) for $p = 2$ (resp. $p > 2$) is understood to be the unit.

The mod p cohomology $H^*(X; \mathbb{F}_p)$ of a space or of a spectrum X will be denoted by H^*X, and the reduced mod p cohomology will be denoted by $\tilde{H}^*X$. Here is the fundamental property of $\mathcal{A}$ (see [SE])

Theorem 1.1.1 (Steenrod, Adem). *For any space (or spectrum) X, H^*X is in a natural way a graded $\mathcal{A}$-module.*

Classically, β (Sq^1 if $p = 2$) acts as the Bockstein homomorphism associated to the sequence $0 \to \mathbb{Z}/p \to \mathbb{Z}/p^2 \to \mathbb{Z}/p \to 0$. N. E. Steenrod constructed the operations Sq^i and the operations P^i, and J. Adem showed that the generators of the ideal of relations act trivially on the mod p cohomology of any space.

The next theorem is a consequence of the computation by H. Cartan and J.-P. Serre (see [C] [S1]) of the cohomology of the Eilenberg-MacLane spaces

Theorem 1.1.2. *The Steenrod algebra is the algebra of all natural stable transformations of mod p cohomology.*

Here 'stable' means 'commuting with suspension' .

1.2. Generators for the Steenrod algebra

First, we describe multiplicative generators for the Steenrod algebra.

Proposition 1.2.1. *The operations Sq^{2^h}, $h \geq 0$, for $p = 2$ constitute a system of multiplicative generators for $\mathcal{A}$; so do the operations β and P^{p^h}, $h \geq 0$ for $p > 2$.*

This is implied by the following proposition which we shall use later.

Proposition 1.2.2. *The operation Sq^m (resp. P^m) for $p = 2$ (resp. $p > 2$) belongs to the right ideal of $\mathcal{A}$ generated by those Sq^{2^t} (resp. P^{p^t}) such that $t \geq 0$ and 2^t (resp. p^t) divides m.*

This is proved using the Adem relations. For p odd write m as $p^\ell + p^\ell a$ where $a = p\alpha + \rho$, $0 \le \rho \le p - 2$. Then

$$P^{p^\ell} P^{p^\ell a} = \sum_0^{p^{\ell-1}} (-1)^{p^\ell + t} \binom{(p-1)(p^\ell a - t) - 1}{p^\ell - pt} P^{m-t} P^t .$$

The coefficient $\varepsilon = \binom{(p-1)p^\ell a - 1}{p^\ell}$ is easily checked to be non-zero mod p. Therefore,

$$P^{p^\ell} P^{p^\ell a} = \varepsilon P^m + \sum_1^{p^{\ell-1}} (-1)^{p^\ell + t} \binom{(p-1)(p^\ell a - t) - 1}{p^\ell - pt} P^{m-t} P^t .$$

Now, for t between 1 and $p^{\ell-1}$, the integer $m - t$ is not divisible by p^ℓ. Therefore we can apply an induction hypothesis and assume that these operations P^{m-t} are in the right ideal generated by the operations P^{p^ℓ} for $t \le \ell - 1$. The result follows similarly for $p = 2$.

In fact the system of generators that we have described is a minimal one. This is proved by looking at the action of the Steenrod algebra on $H^*B\mathbb{Z}/p$ (see 1.4). The operation P^{p^t} is indecomposable (it is not a sum of products of elements of strictly lower degree). Indeed it acts non-trivially on $H^{2p^t}B\mathbb{Z}/p$ whereas decomposable elements act trivially. A similar argument works when $p = 2$. We refer to [SE] for further details.

We now describe an additive basis for the Steenrod algebra. In order to do that we need to describe certain monomial elements in $\mathcal{A}$.

Let p be 2. For a sequence of integers $I = (i_1, \ldots, i_n)$, let Sq^I denote $\mathrm{Sq}^{i_1} \ldots \mathrm{Sq}^{i_n}$. The sequence I is said to be **admissible** if $i_h \ge 2i_{h+1}$ for all $h \ge 1$, $(i_{n+1} = 0)$.

Let $p > 2$. For a sequence of integers $I = (\varepsilon_0, i_1, \varepsilon_1, \ldots, i_n, \varepsilon_n)$, where the ε_k are 0 or 1, let P^I denote $\beta^{\varepsilon_0} P^{i_1} \beta^{\varepsilon_1} \ldots \beta^{\varepsilon_n}$. The sequence I is said to be **admissible** if $i_h \ge pi_{h+1} + \varepsilon_h$ for all $h \ge 1$, $(i_{n+1} = 0)$.

The operations Sq^I (resp. P^I) with I admissible are called admissible monomials.

Proposition 1.2.3. *The admissible monomials Sq^I (resp. P^I) form a basis for $\mathcal{A}$.*

It is a consequence of the Adem relations that the operations Sq^I (resp. the operations P^I), I admissible, span the graded vector space $\mathcal{A}$. We prove it now. Let $I = (i_1, \ldots, i_n)$ (resp. $I = (\varepsilon_0, i_1, \varepsilon_1, \ldots, i_n, \varepsilon_n)$ be an admissible sequence. Its moment is defined to be $i_1 + 2i_2 + \ldots + ni_n$ (resp. $i_1 + \varepsilon_1 + 2(i_2 + \varepsilon_2) + \ldots$. If I is not admissible there exists h, $1 \le h \le n - 1$ such that $i_h < 2i_{h+1}$. Using the Adem relations one gets:

$$\mathrm{Sq}^I = \sum_0^{[i_h/2]} \varepsilon_t \, \mathrm{Sq}^{I'} \, \mathrm{Sq}^{i_h + i_{h+1} - t} \, \mathrm{Sq}^t \, \mathrm{Sq}^{I''},$$

where $\varepsilon_t \in \mathbb{F}_2$, $0 \le t \le [i_h/2]$, $I' = (i_1, \ldots, i_{h-1})$ and $I'' = (i_{h+2}, \ldots, i_n)$. The moment of any sequence occurring on the right (i.e. $(I', i_h + i_{h+1} - t, t, I'')$, $0 \le t \le [i_h/2]$) is strictly lower than the moment of I. Thus the result follows clearly by induction on the moment. The case of an odd prime is proved in the same way.

In order to show that the admissible monomials form a basis it remains to show that they are linearly independent. To do that one looks at their action on $H^*(B\mathbb{Z}/p^{\oplus k}) \cong (H^*B\mathbb{Z}/p)^{\otimes k}$ (see 1.4 in [SE] for details). If $p = 2$, one shows that the elements $\mathrm{Sq}^I(u^{\otimes k})$, with I admissible and $|\mathrm{Sq}^I| \le k$, are linearly independent. If p is an odd prime one considers the action of the operations P^I on the element $tx \otimes \ldots \otimes tx$ (again see 1.4 for the notation) of $H^*(B\mathbb{Z}/p^{\oplus k})$. The result follows.

1.3. The instability condition

The mod p cohomology of a space X has, as $\mathcal{A}$-module, a certain property called *instability*:

— if $x \in H^*X$ and $i > |x|$, then $\mathrm{Sq}^i x = 0$, for $p = 2$;

— if $x \in H^*X$ and $e + 2i > |x|$, $e = 0, 1$, then $\beta^e P^i x = 0$, for $p > 2$.

Here $|x|$ denotes the degree of x.

Definition 1.3.1.   An $\mathcal{A}$-module M is unstable if it satisfies the preceding property.

In particular, this implies that an unstable $\mathcal{A}$-module M is trivial in negative degrees (recall that one identifies Sq^0, resp. P^0, with the identity operator).

Hence, the cohomology of a spectrum, which need not be non-trivial in negative degrees, is not in general an unstable $\mathcal{A}$-module.

For convenience, we shall abbreviate 'unstable $\mathcal{A}$-module' to 'unstable module' most of the time.

We shall denote by $\mathcal{M}$ the category of $\mathbb{Z}$-graded $\mathcal{A}$-modules, morphisms being $\mathcal{A}$-linear map of degree zero. We shall denote by $\mathcal{U}$ the full subcategory whose objects are unstable $\mathcal{A}$-modules.

Before discussing examples, let us introduce unstable $\mathcal{A}$-algebras.

1.4. Unstable $\mathcal{A}$-algebras

The mod p cohomology of a space X is also, in a natural way, a graded commutative, unital, $\mathbb{F}_p$-algebra. The algebra structure is related to the $\mathcal{A}$-module structure by two properties:

$$(\mathcal{K}1) \quad \begin{cases} \mathrm{Sq}^i(xy) = \sum_{k+\ell=i} \mathrm{Sq}^k x\, \mathrm{Sq}^\ell y \\ \qquad \text{for any } x \text{ and } y \text{ in } H^*X \text{ if } p = 2, \\ P^i(xy) = \sum_{k+\ell=i} P^k x\, P^\ell y, \\ \beta(xy) = (\beta x)\, y + (-1)^{|x|} x\, \beta y \\ \qquad \text{for any } x \text{ and } y \text{ in } H^*X \text{ if } p > 2. \end{cases}$$

The axiom $(\mathcal{K}1)$ is known as the 'Cartan formula'.

$$(\mathcal{K}2) \quad \begin{cases} \mathrm{Sq}^{|x|} x = x^2 & \text{for any } x \text{ in } H^*X \text{ if } p = 2, \\ P^{|x|/2}x = x^p & \text{for any } x \text{ of even degree in } H^*X \text{ if } p > 2. \end{cases}$$

This leads to

Definition 1.4.1. An unstable $\mathcal{A}$-algebra K is an unstable $\mathcal{A}$-module provided with maps $\mu : K \otimes K \to K$ and $\eta : \mathbb{F}_p \to K$ which determine a commutative, unital, $\mathbb{F}_p$-algebra structure on K and such that properties $(\mathcal{K}1)$ and $(\mathcal{K}2)$ hold.

We shall denote by $\mathcal{K}$ the category of unstable $\mathcal{A}$-algebras, morphisms being $\mathcal{A}$-linear algebra maps of degree zero. For convenience, we shall often abbreviate 'unstable $\mathcal{A}$-algebra' to 'unstable algebra'.

The axiom $(\mathcal{K}1)$ can be reformulated as follows. There is an algebra map δ (diagonal) from $\mathcal{A}$ in $\mathcal{A} \otimes \mathcal{A}$ such that

$$\delta(\mathrm{Sq}^i) = \sum_{k+\ell=i} \mathrm{Sq}^k \otimes \mathrm{Sq}^\ell \quad \text{if} \quad p = 2;$$

$$\begin{cases} \delta(\beta) &= \beta \otimes 1 + 1 \otimes \beta, \\ \delta(P^i) &= \sum_{k+\ell=i} P^k \otimes P^\ell \quad \text{if} \quad p > 2. \end{cases}$$

This map determines a co-commutative Hopf algebra structure on $\mathcal{A}$, and it can be used to provide the tensor product $M \otimes N$ of two $\mathcal{A}$-modules M and N with an $\mathcal{A}$-module structure, this structure being determined by the formula

$$(\theta \otimes \theta')(m \otimes n) = (-1)^{|\theta'||m|} \theta m \otimes \theta' m'$$

for all $\theta, \theta' \in \mathcal{A}$, $m \in M$ and $n \in N$. Then $M \otimes N$ is an $\mathcal{A}$-module by restriction via δ. Axiom $(\mathcal{K}1)$ is equivalent to the $\mathcal{A}$-linearity of the map $\mu : K \otimes K \to K$. The structure of the dual of $\mathcal{A}$, as a commutative Hopf algebra, which was determined by Milnor [**Mn**] is given in (1.10).

1.5. Notation and basic example

As an example of an unstable algebra, recall the structure of the mod p cohomology of the space $B\mathbb{Z}/p$.

The mod 2 cohomology $H^*B\mathbb{Z}/2$ is the polynomial algebra $\mathbb{F}_2[u]$ on one generator u of degree 1. The action of $\mathcal{A}$ is completely determined by axioms $(\mathcal{K}1)$ and $(\mathcal{K}2)$ and one finds that $\mathrm{Sq}^i u^n = \binom{n}{i} u^{n+i}$; if $p > 2$, $H^*B\mathbb{Z}/p$ is the tensor product $E(t) \otimes \mathbb{F}_p[x]$ of an exterior algebra on one generator t of degree 1 and a polynomial algebra on one generator x of degree 2. The action of $\mathcal{A}$ is determined by axioms $(\mathcal{K}1)$ and $(\mathcal{K}2)$ and the fact that β is the Bockstein homomorphism. We obtain

$$\beta t = x \quad , \quad P^i x^n = \binom{n}{i} x^{n+i(p-1)}.$$

From now on (unless otherwise specified), the letter V will denote an elementary abelian p-group, in other words a finite dimensional $\mathbb{F}_p$-vector space. The mod p cohomology of BV will be denoted by H^*V, and the reduced mod p cohomology of BV will be denoted by $\tilde{H}^*V$. When $p = 2$, H^*V identifies with the symmetric algebra on V^* concentrated in degree 1, $S^*(V^*)$. When $p > 2$ H^*V identifies with the tensor product of the exterior algebra on V^* concentrated in degree 1, $E(V^*)$, and of the symmetric algebra on another copy of V^* concentrated in degree 2, $S^*(V^*)$.

A map of unstable $\mathcal{A}$-algebras from H^*W in H^*V is determined by its behavior in degree 1. In degree 1, it identifies with a linear map $\varphi : V \to W$. It follows easily that the map

$$\text{Hom}(V, W) \to \text{Hom}_{\mathcal{K}}(H^*W, H^*V), \quad \varphi \mapsto \varphi^*,$$

is a bijection.

1.6. Free objects in the category $\mathcal{U}$

The category $\mathcal{U}$ is obviously an abelian category. It has enough projective objects. This is implied by

Proposition 1.6.1 [SE][MP]. *There is, up to isomorphism, a unique unstable $\mathcal{A}$-module $F(n)$ with a class ι_n of degree n such that the natural transformation $f \mapsto f(\iota_n)$ from $\text{Hom}_{\mathcal{U}}(F(n), M)$ in M^n is an equivalence of functors.*

The functor $M \mapsto M^n$ being right exact, $F(n)$ is projective, so it deserves to be called the free unstable module on one generator of degree n.

We need a definition. The **excess** of an admissible sequence I is defined to be

$$(i_1 - 2i_2) + (i_2 - 2i_3) + \ldots + (i_{n-1} - 2i_n) + i_n \text{ if } p = 2,$$
$$2(i_1 - pi_2) + 2(i_2 - pi_3) + \ldots + 2i_n + \varepsilon_0 - \varepsilon_1 - \ldots - \varepsilon_n \text{ if } p > 2,$$

and it is denoted by $e(I)$. Note that, if $p > 2$, an admissible sequence I such that $e(I) \leq n$ contains at most n entries $\varepsilon_i = 1$ because $e(I) = \varepsilon_0 + \ldots + \varepsilon_n + 2(i_1 - pi_2 - \varepsilon_1) + \ldots + 2(i_n - \varepsilon_n)$.

Recall the definition of the suspension functor $\Sigma : \mathcal{M} \to \mathcal{M}$. Given an $\mathcal{A}$-module M, ΣM is defined by $(\Sigma M)^n \cong M^{n-1}$. The $\mathcal{A}$-action is given by $\theta(\Sigma m) = (-1)^{|\theta|} \Sigma \theta m$ for all $m \in M$, $\theta \in \mathcal{A}$.

We can now describe the unstable $\mathcal{A}$-module $F(n)$:

Proposition 1.6.2. *The unstable $\mathcal{A}$-module $F(n)$ is isomorphic to* $\Sigma^n (\mathcal{A}/(P^I, I$ *is admissible and* $e(I) > n))$.

Hence this module has as an $\mathbb{F}_p$-basis the elements $\Sigma^n P^I$, with I admissible and $e(I) \leq n$. We shall denote $\Sigma^n P^I$ by $P^I \iota_n$.

Proof of 1.6.2 (Compare with [SE]).

Essentially we have to prove that the $\mathbb{F}_p$-span of the elements P^I with I admissible and $e(I) > n$ is a sub $\mathcal{A}$-module of $\mathcal{A}$. Propositions 1.6.2 and 1.6.1 follow directly.

Assume for simplicity that $p = 2$. For any sequence $I = (i_1, \dots, i_n)$ denote by $|I|$ the sum $i_1 + \dots + i_n$. Extend the definition of the excess to all sequences: $e(I) = i_1 - i_2 - \dots - i_n$. Given an admissible sequence I and an integer $k \geq 0$ we have to show that the operation $Sq^k Sq^I$ decomposes as a sum of operations Sq^J with J admissible and $e(J) \geq e(I)$.

Let $I = (i_1, \dots, i_n)$ be an admissible sequence and $k \geq 0$ be an integer. If $k \geq 2i_1$ there is nothing to do because $(k, i_1, \dots, i_n)$ is admissible and the excess of this sequence is greater than $e(I)$.

If $k < 2i_1$ then using the Adem relations one gets:

$$Sq^k Sq^I = \sum_0^{[k/2]} \varepsilon_t \, Sq^{k+i_1-t} Sq^t Sq^{I'},$$

where $\varepsilon_t \in \mathbb{F}_2$, $I' = (i_2, \dots, i_n)$ and $0 \leq t \leq [k/2]$. The excess of any sequence $(k + i_1 - t, t, i_2, \dots, i_n)$, $0 \leq t \leq [k/2]$, is easily checked to be greater than $e(I)$. Let I_t be the sequence $(t, i_2, \dots, i_n)$. The operation Sq^{I_t} can be written as $\sum_u Sq^{J_{t,u}}$ with $J_{t,u}$ admissible, $|J_{t,u}| = t + i_2 + \dots + i_n$. The excess of $(k + i_1 - t, J_{t,u})$, as it is equal to the excess of $(k + i_1 - t, t, i_2, \dots, i_n)$, is greater than $e(I)$. If all of these sequences are admissible we are done. Otherwise one applies an Adem relation to $Sq^{k+i_1-t} Sq^{J_{t,u}}$ as above.

This increases strictly the first index in the sequences that appear. Therefore, as this index is bounded by $k + |I|$, this process has to stop after a finite number of steps. We are done. The case of an odd prime is left to the reader.

Examples.

1) If $p = 2$, $F(1)$ can be identified with the sub $\mathcal{A}$-module of $H^*\mathbb{Z}/2$ generated by the class u. The set $\{u, u^2, u^4, \ldots\}$ is a basis for $F(1)$.

2) If $p > 2$, $F(1)$ can be identified with the sub $\mathcal{A}$-module of $H^*\mathbb{Z}/p$ generated by the class t. The set $\{t, x, x^p, x^{p^2}, \ldots\}$ is a basis for $F(1)$.

It is easy to check that if M and N are unstable, $M \otimes N$ equipped with the diagonal action is unstable. Consider the unstable module $F(1)^{\otimes n}$; it is of dimension 1 in degree n. Therefore, up to a scalar multiple, there is one non-trivial map $\omega_n : F(n) \to F(1)^{\otimes n}$. The symmetric group $\mathfrak{S}_n$ acts on $F(1)^{\otimes n}$ by permutation of the coordinates. If $p = 2$ the non-trivial class of degree n of $F(1)^{\otimes n}$ is $\mathfrak{S}_n$-invariant. Therefore, ω_n takes values in the invariants under the action of $\mathfrak{S}_n$ which are denoted $(F(1)^{\otimes n})^{\mathfrak{S}_n}$.

Proposition 1.6.3. *If $p = 2$, the map ω_n is an isomorphism from $F(n)$ onto $(F(1)^{\otimes n})^{\mathfrak{S}_n}$.*

The proof that we shall give depends on the functor Φ which is going to be defined in the next section. It will be given in (1.9). The reader can also give a direct proof based for example on Milnor's coaction (see the appendix of the Chapter).

Such a result cannot be true for $p > 2$. This is because, for example, the operation P^1 is always trivial on $t \otimes \ldots \otimes t$ but not on ι_n if $n \geq 2$.

However, it is still possible to make a statement for odd primes. Restricting attention to the full subcategory $\mathcal{U}'$ of $\mathcal{U}$ consisting of unstable modules which are non-trivial only in even degrees, one shows that $\mathcal{U}'$ has projective objects $F'(2n)$ defined in a similar way as the unstable $\mathcal{A}$-modules $F(n)$. Moreover, the elements $P^I i_{2n}$ where I is an admissible sequence with $e(I) \leq 2n$ and $\varepsilon_i = 0$ for all

i form a $\mathbb{F}_p$-basis for $F'(2n)$. With this definition, (1.6.3) extends immediately to

(1.6.4) $F'(2n) \cong (F'(2)^{\otimes n})^{\mathfrak{S}_n}$.

We shall use the category $\mathcal{U}'$ in various places.

1.7. Instability and the Adem relations

The Adem relations are delicate to use. However some of them simplify in an unstable module M. Let us introduce the operation Sq_0 (resp. P_0) for $p = 2$ (resp. $p > 2$) which is defined as follows:

$$\begin{aligned} Sq_0 x &= Sq^{|x|} x & \text{if } p = 2\,; \\ P_0 x &= P^{|x|/2}x & \text{if } |x| \equiv 0\,(2) \quad \text{and} \\ P_0 x &= \beta P^{(|x|-1)/2}x & \text{if } |x| \equiv 1\,(2)\,, \text{ if } p > 2\,. \end{aligned}$$

It is an exercise to check that for any element x in an unstable module M and any $i \geq 0$,

(1.7.1) $Sq^i Sq_0 x = \begin{cases} Sq_0 Sq^{i/2} x & \text{if } i \equiv 0\,(2)\,, \\ 0 & \text{otherwise}\,. \end{cases}$

(1.7.1*) $\begin{cases} P^i P_0 x = P_0 P^{i/p} x & \text{if } i \equiv 0\,(p)\,, \\ P^i P_0 x = P_0 \beta P^{(i-1)/p} x & \text{if } i \equiv 1\,(p) \text{ and } |x| \equiv 1\,(2)\,, \\ P^i P_0 x = 0 \text{ otherwise}\,, \\ \beta P_0 x = 0\,. \end{cases}$

Let us prove for example (1.7.1). The only interesting case is when $i < 2|x|$. One can apply the Adem relations and one gets:

$$Sq^i Sq^{|x|} x = \sum_0^{[i/2]} \binom{|x| - t - 1}{i - 2t} Sq^{|x|+i-t} Sq^t x\,.$$

But $Sq^{|x|+i-t}(Sq^t x)$ is trivial as soon as $|x| + i - t > t + |x|$. Therefore the term on the right reduces to zero if i is odd and to $Sq^{|x|+i/2} Sq^{i/2} x$ if i is even.

This leads us to introduce the following functor Φ from $\mathcal{U}$ to itself. Given $M \in \mathcal{U}$, define ΦM for p odd by

$$(1.7.2) \qquad (\Phi M)^n = \begin{cases} M^{n/p} & \text{if } n \equiv 0 \ (2p) \\ M^{(n-2)/p+1} & \text{if } n \equiv 2 \ (2p) \\ 0 & \text{otherwise}, \end{cases}$$

the action of the Steenrod algebra is given by

$$P^i \Phi x = \Phi P^{i/p} x \quad \text{if } p \mid i$$
$$P^i \Phi x = \Phi \beta P^{i-1/p} x \quad \text{if } p \mid i-1 \quad \text{and} \quad |x| \equiv 1 \ (2)$$
$$P^i \Phi x = 0 \quad \text{otherwise, and}$$
$$\beta \Phi x = 0,$$

where Φx denotes the element of ΦM corresponding to x in M.

If $p = 2$, these formulas simplify dramatically

$$(\Phi M)^n \cong M^{n/2}, \qquad \mathrm{Sq}^i \, \Phi x = \Phi \, \mathrm{Sq}^{i/2} x,$$

where $M^{n/2}$ (resp. $\mathrm{Sq}^{i/2}$) is trivial if $n/2$ (resp. $i/2$) is not an integer. See **[MP]**, **[Li]** and **[LZ1]** for more details. One checks that ΦM is indeed an object in $\mathcal{U}$. The relations (1.7.1*) and (1.7.1) express the fact that the map $\lambda_M : \Phi M \to M$, $\Phi x \mapsto P_0 x$ (resp. $\Phi x \mapsto \mathrm{Sq}_0 x$) is $\mathcal{A}$-linear.

We are interested in $\mathrm{Ker} \, \lambda_M$ and $\mathrm{Coker} \, \lambda_M$. The following case is especially important. Let $\Sigma\iota_{n-1} : F(n) \to \Sigma F(n-1)$ be the map which sends ι_n to $\Sigma\iota_{n-1}$

Proposition 1.7.3. *The following is an exact sequence of $\mathcal{A}$-modules*

$$0 \to \Phi F(n) \xrightarrow{\lambda_{F(n)}} F(n) \xrightarrow{\Sigma\iota_{n-1}} \Sigma F(n-1) \to 0.$$

This is proved by using the basis of $F(n)$ given in (1.6). First one checks that $\Phi F(n)$ identifies as a submodule of $F(n)$ with the $\mathbb{F}_p$-span of those $P^I \iota_n$ where $e(I) = n$, i.e. all classes of the form $P_0 x$. Therefore $\lambda_{F(n)}$ is injective. Next, as a class of the form $P_0 x$ (resp. $\mathrm{Sq}_0 x$) is always trivial in a suspension, by the instability condition, the composite $\Sigma\iota_{n-1} \circ \lambda_{F(n)}$ is trivial. It remains to prove exactness at $F(n)$ and $\Sigma F(n-1)$. These follow from the description given in (1.6).

Observe now that Ker λ_M and Coker λ_M are, by the very definition, suspensions. This leads us to define functors Ω and Ω_1 from $\mathcal{U}$ to itself by

(1.7.4)
$$\begin{cases} \mathrm{Ker}\ \lambda_M = \Sigma\Omega_1 M\,, \\ \mathrm{Coker}\ \lambda_M = \Sigma\Omega M\,. \end{cases}$$

In other words we have an exact sequence

$$0 \to \Sigma\Omega_1 M \to \Phi M \to M \to \Sigma\Omega M \to 0\,.$$

Proposition 1.7.5. *The functor Ω is the left adjoint of the suspension functor Σ. The functor Ω_1 is the first left derived functor of Ω, and all higher derived functors are trivial.*

Note that, as we are working in cohomology, the functor Σ has a left adjoint, instead of having a right one as in the topological setting. We refer to **[MP]**, **[LZ1]** and **[Si]** for more details and related results concerning derived functors of the iterates of Ω.

The proofs are 'routine': one considers a free resolution for M

$$\ldots \to P_1 \to P_0 \to M \to 0\,.$$

where the P_i are direct sums of $F(n)$'s. Then one looks at the diagram

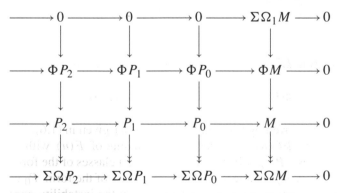

Here the two middle rows are exact (as Φ is obviously exact), all columns are exact, and the top right vertical arrow is injective. The result follows by inspection.

1.8. The category $\mathcal{U}$ is locally noetherian

Theorem 1.8.1. *Let M be an unstable $\mathcal{A}$-module which has finitely many generators as an $\mathcal{A}$-module. Then any sub $\mathcal{A}$-module of M has also finitely many generators as $\mathcal{A}$-module.*

This implies that the category $\mathcal{U}$ is locally noetherian (using P. Gabriel's terminology, see [**Gb**], and also [**Po**]). Recall that an object in an abelian category is noetherian if it has the ascending chain property. An object has the ascending chain property if any ascending sequence of subobjects stabilizes. A category is locally noetherian if it has a set of noetherian generators. Recall that a set of objects $\{M_\alpha\}$, $\alpha \in A$, in an abelian category is called a set of generators, if for any object M in the category, there is an epimorphism from a certain direct sum of the M_α's onto M. Following Gabriel we assume moreover, as part of the definition, that the category has exact filtered colimits. Theorem 1.8.1 implies that the generators $F(n)$ of the category $\mathcal{U}$ are noetherian.

This has been proved by W. Massey and F. Peterson for $p = 2$, see [**MP**], Steenrod gave a proof (unpublished) for $p > 2$. Lannes and Zarati published a proof for all p in [**LZ1**]. We reproduce these proofs here with minor changes.

Proof. Standard arguments show that it is enough to prove that any sub-module of $F(n)$ has finitely many generators over $\mathcal{A}$. Obviously, one can assume that $n > 0$.

1) First we treat the case $p = 2$. The idea is to proceed by induction using (1.7.3). Let $M \subset F(n)$, and define M_k to be the sub $\mathcal{A}$-module of $F(n)$ of those x such that $\mathrm{Sq}_0^k x$ is in M. We first show that the increasing sequence

$$M = M_0 \subset M_1 \subset \ldots \subset M_k \subset \ldots$$

stabilizes. One considers the diagram

(1.8.2)

$$
\begin{array}{ccccccccc}
0 & \longrightarrow & \Phi M_{k+1} & \longrightarrow & M_k & \longrightarrow & \Sigma M_k' & \longrightarrow & 0 \\
& & \downarrow & & \downarrow & & \downarrow & & \\
0 & \longrightarrow & \Phi M_{k+2} & \longrightarrow & M_{k+1} & \longrightarrow & \Sigma M_{k+1}' & \longrightarrow & 0
\end{array}
$$

where the rows are exact (exercise!) and where $\Sigma M_k'$ is the image of M_k in $\Sigma F(n-1)$. The sequence $\ldots \subset M_k' \subset M_{k+1}' \subset \ldots$ stabilizes at a certain integer k_0 by induction on n. The sequence of the submodules M_k also has to stabilize at k_0. If not, there would exist a non-zero $x_1 \in M_{k_0+1}$, $|x_1| > 0$, which is sent to zero in $\Sigma M_{k_0+1}'$ and which is not in the image of M_{k_0}. Hence x_1 is equal to $\Phi x_1'$ for some x_1' in M_{k_0+2} but not in M_{k_0+1}. Consequently there exists a non-zero $x_2 \in M_{k_0+2}$ of degree $|x_1|/2$ which is sent to zero in $\Sigma M_{k_0+2}'$, and which is not in the image of M_{k_0+1}. An obvious induction leads to a contradiction. It follows that M_{k_0}' is isomorphic to ΩM_{k_0}.

Observe now that M_{k_0} is connected ($M_{k_0}^0 = 0$) because $n > 0$.

Lemma 1.8.3. *Let M be a connected ($M^0 = 0$) unstable $\mathcal{A}$-module. If ΩM is finitely generated over $\mathcal{A}$, so is M.*

Choose elements $a_1, \ldots, a_t$ in M which are liftings of $\mathcal{A}$-generators of $\Sigma \Omega M$. Then the map $f : \oplus_1^t F(|a_i|) \to M$ is surjective. Indeed observe that $\Omega(\operatorname{Coker} f) = 0$ because $\Omega(f)$ is surjective and Ω is right exact. Then, Coker f being connected, one concludes easily that Coker $f = 0$. Therefore M_{k_0} is finitely generated as an $\mathcal{A}$-module. It remains to use the following exact sequence

$$0 \to \Phi M_k \to M_{k-1} \to \Sigma M_k' \to 0$$

to show that if M_k is finitely generated so is M_{k-1}. One uses the following facts:

— M_k' is finitely generated (induction hypothesis);

— if M is finitely generated so is ΦM.

2) The case $p > 2$. The above proof works with $\mathcal{U}$ replaced by $\mathcal{U}'$ when $p > 2$.

Then one observes that *the span over $\mathbb{F}_p$ of those $P^I \iota_n \in F(n)$ such that $\Sigma \varepsilon_i \geq k$ is a sub $\mathcal{A}$-module which we shall denote by $F(n)_k$.* Note that $F(1)_1 \cong F'(2)$ (1.6.4). There is a filtration:

$$F(n) = F(n)_0 \supset F(n)_1 \supset \ldots \supset F(n)_n \supset F(n)_{n+1} = 0.$$

The quotient $F(n)_k/F(n)_{k+1}$ is finitely generated and is the direct sum of an object of $\mathcal{U}'$ and of the suspension of an object in $\mathcal{U}'$. Only the first claim needs some explanation. It is proved using the Adem relations involving β. One shows that it is generated by the image of classes $P^I \iota_n \in F(n)$ *such that* $\Sigma \varepsilon_i = k$, and such that $\varepsilon_i = 0$ for $i > k$.

From this point, the result follows by classical arguments.

(1.8.4) Exercise. Let p be an odd prime, M be an unstable module with generators $a_1, \ldots, a_k$. Describe generators for ΦM.

Corollary 1.8.5. *The unstable $\mathcal{A}$-module $F(p) \otimes F(q)$ is finitely generated for all p and q. Therefore, the tensor product of two finitely generated unstable $\mathcal{A}$-modules is finitely generated.*

The first assertion is going to be proved in (1.9) using (1.6.3), (1.8.1) and (1.8.3). The second one follows easily. If M and N are finitely generated they are quotient of certain finite direct sums $\bigoplus_1^m F(a_i)$, and $\bigoplus_1^n F(b_j)$. Therefore, $M \otimes N$ is a quotient of the finite direct sum $\bigoplus_{i,j} F(a_i) \otimes F(b_j)$ which is finitely generated.

1.9. Proof of Proposition 1.6.3 and of Corollary 1.8.5

Proof of 1.6.3 (compare with [LZ2]).   We proceed by induction over n. Observe that Φ commutes with tensor product and with invariants inside $F(1)^{\otimes n}$. Consider the commutative diagram

$$
\begin{array}{ccccccccc}
0 & \longrightarrow & \Phi(F(n)) & \longrightarrow & F(n) & \longrightarrow & \Sigma F(n-1) & \longrightarrow & 0 \\
& & \downarrow{\scriptstyle \Phi\omega_n} & & \downarrow{\scriptstyle \omega_n} & & \downarrow{\scriptstyle \Sigma\omega_{n-1}} & & \\
0 & \rightarrow & (\Phi F(1)^{\otimes n})^{\mathfrak{S}_n} & \rightarrow & (F(1)^{\otimes n})^{\mathfrak{S}_n} & \overset{\pi}{\rightarrow} & \Sigma(F(1)^{\otimes n})^{\mathfrak{S}_{n-1}} & \rightarrow & 0
\end{array}
$$

Only π needs to be defined; it is the restriction of the map

$$ p \otimes \mathrm{Id}_{n-1} : F(1) \otimes F(1)^{\otimes n-1} \rightarrow \Sigma \mathbb{F}_2 \otimes F(1)^{\otimes n-1} , $$

where $p : F(1) \rightarrow \Sigma \mathbb{F}_2$ is the non-trivial map. One checks that the bottom row is exact (this is not true before taking invariants). By hypothesis $\Sigma \omega_{n-1}$ is an isomorphism. The snake lemma implies that the maps $\Phi(\mathrm{Ker}\,\omega_n) \rightarrow \mathrm{Ker}\,\omega_n$ and $\Phi(\mathrm{Coker}\,\omega_n) \rightarrow \mathrm{Coker}\,\omega_n$ are isomorphisms. But then $\mathrm{Ker}\,\omega_n$ and $\mathrm{Coker}\,\omega_n$ must be concentrated in degree zero. But, if $n \geq 1$, they are trivial in degree zero. We are done.

Proof of 1.8.5. Let $p = 2$. By Theorem 1.8.1 it is enough to prove that $F(1)^{\otimes p+q}$ is finitely generated. Indeed by Proposition 1.6.3 $F(p) \otimes F(q)$ is a sub $\mathcal{A}$-module of $F(1)^{\otimes p+q}$.

There is an exact sequence

$$0 \to F(1)^{\otimes n} \to F(1)^{\otimes n} \xrightarrow{\pi'} \bigoplus_1^n \Sigma F(1)^{\otimes(n-1)} .$$

where the map π' is the sum of the maps

$$F(1)^{\otimes n} \xrightarrow{\cong} F(1)^{\otimes(n-i-1)} \otimes F(1) \otimes F(1)^{\otimes i}$$
$$\to F(1)^{\otimes(n-i-1)} \otimes \Sigma \mathbb{F}_2 \otimes F(1)^{\otimes i} \xrightarrow{\cong} \Sigma F(1)^{\otimes(n-1)} .$$

Thus the unstable A-module $\Omega F(1)^{\otimes n}$ embeds in $\bigoplus_1^n F(1)^{\otimes(n-1)}$. We can apply an induction hypothesis to conclude that it is finitely generated. Then the result follows from Lemma 1.8.3. The case of an odd prime is treated by using $\mathcal{U}'$.

Corollary 1.8.5 has surely been known for a long time by various people. It is implicit in [**L1**]. N. Kuhn and T. Kashiwabara have also given proofs.

1.10. Appendix on Milnor's dual $\mathcal{A}_*$ of the Steenrod algebra and on Milnor's derivations

First, we recall the structure of the dual algebra $\mathcal{A}_*$ of $\mathcal{A}$. It was computed by Milnor in [**Mn**] (see also [**SE**]).

Proposition 1.10.1. *There is a (unique) algebra map* $\delta : A \to A \otimes A$ *such that for all* $i \geq 0$

$$\delta(\mathrm{Sq}^i) = \Sigma_{j+k=i} \mathrm{Sq}^j \otimes \mathrm{Sq}^k \quad \text{if } p = 2;$$
$$\delta(P^i) = \Sigma_{j+k=i} P^j \otimes P^k, \text{ and } \delta(\beta) = \beta \otimes 1 + 1 \otimes \beta \quad \text{if } p > 2.$$

This can be proved by using the Kunneth theorem and the Cartan formula. Consequently the Steenrod algebra is a co-commutative Hopf algebra. Its dual $\mathcal{A}_*$ is a commutative Hopf algebra.

Theorem 1.10.2. *The algebra* $\mathcal{A}_*$ *is*

(i) *polynomial on generators* ξ_i, $i \geq 1$, *of degree* $2^i - 1$ *if* $p = 2$;

(ii) *the tensor product of a polynomial algebra on generators* ξ_i, $i \geq 1$, *of degree* $2(p^i - 1)$ *by an exterior algebra on generators* τ_i, $i \geq 0$, *of degree* $2p^i - 1$ *if* $p > 2$.

Moreover, the coproduct δ *is given by the formulas*

(iii) $\delta(\xi_k) = \sum_{0 \leq i \leq k} \xi_{k-i}^{2^i} \otimes \xi_i$ *if* $p = 2$;

(iv) $\delta(\xi_k) = \sum_{0 \leq i \leq k} \xi_{k-i}^{p^i} \otimes \xi_i$ *and*

$\delta(\tau_k) = \tau_k \otimes 1 + \sum_{0 \leq i \leq k} \xi_{k-i}^{p^i} \otimes \tau_i$, *if* $p > 2$.

In these formulas, ξ_0 is understood to be 1.

If I is a multi-index $(i_1, \ldots, i_r)$, the monomial (in $\mathcal{A}_*$) $\xi_1^{i_1} \ldots \xi_r^{i_r}$ is written as ξ^I and $\lambda(x)$ is written as a formal sum $\sum_I x_I \otimes \xi^I$.

Certain operations introduced by Milnor are very useful when working with modules over the Steenrod algebra. We recall here briefly their properties. Recall first Milnor's coaction (see [**Mn**])

$$\lambda : M \to M \hat{\otimes} \mathcal{A}_*$$

for an $\mathcal{A}$-module M. The completed tensor product $M \hat{\otimes} \mathcal{A}_*$ is the graded $\mathbb{F}_2$-vector space which is given by

$$(M \hat{\otimes} \mathcal{A}_*)^n = \prod_{\ell - k = n} M^\ell \otimes \mathcal{A}_k .$$

We write $\lambda(x)$ as a formal sum:

$$\lambda(x) = \Sigma_I x_I \otimes \xi^I .$$

The map $\lambda : M \to M \hat{\otimes} \mathcal{A}_*$ is of degree zero and determined by the following property

— for all $\theta \in \mathcal{A}$, $x \in M$, $\theta x = \sum_I \xi^I(\theta) x_I = \lambda(x)/\theta$.

The coaction λ is given on a tensor product $M \otimes N$ by the following formula

(1.10.3) $$\lambda(x \otimes y) = \sum_I \sum_{J + K = I} (x_J \otimes y_K) \otimes \xi^I ,$$

the sum of multi-indices being taken index by index.

In the sequel, for simplicity, we are going to write down the formulas at the prime 2. However it is easy to extend what follows to odd primes.

The following fundamental example is used by Milnor to define the ξ_i's introduced in (1.10.2). Let u be the generator of $H^1\mathbb{Z}/2$. Then the ξ_i's are defined by

$$(1.10.4) \qquad \lambda(u) = \sum_{i \geq 0} u^{2^i} \otimes \xi_i \quad (\xi_0 = 1) \, .$$

Let us now introduce Milnor's operations Q_i, $i \geq 0$ [**Mn**]. They are dual, with respect to the monomial basis of $\mathcal{A}_*$, to the ξ_{i+1}'s. They commute with each other, and their square is trivial. They can also be defined by

$$Q_0 = \mathrm{Sq}^1 \, , \qquad Q_{i+1} = Q_i \, \mathrm{Sq}^{2^{i+1}} - \mathrm{Sq}^{2^{i+1}} Q_i \, .$$

They are derivations, i.e. $Q_i(x \otimes y) = Q_i x \otimes y + x \otimes Q_i y$ for all x and y in any unstable $\mathcal{A}$-modules M and N. In particular, if u denotes the generator of $H^1\mathbb{Z}/2$, one has $Q_i u^{2\ell} = 0$ and $Q_i u^{2\ell+1} = u^{2\ell+2^{i+1}}$.

More generally one can consider the operations P_t^s that are dual, in the monomial basis of $\mathcal{A}_*$, to the elements $\xi_t^{2^s}$. Their effect is determined by the formula

$$(1.10.5) \qquad \lambda(u^{2^s}) = \sum_{i \geq 0} u^{2^{s+i}} \otimes \xi_i^{2^s} \quad (\xi_0 = 1) \, ,$$

in particular, $P_t^s(u^{2^i}) = u^{2^{s+t}}$ if $i = s$ and $P_t^s(u^{2^i}) = 0$ otherwise. We also note that the operations P_t^s act as derivations on the unstable $\mathcal{A}$-algebra $\Phi^s K$ for any unstable $\mathcal{A}$-algebra K.

One defines similar operations at odd primes. They are dual to the elements $\xi_t^{2^s}$, and also called P_t^s. One has (see 1.5) $P_t^s(t) = 0$, and $P_t^s(x^{2^i}) = x^{2^{s+t}}$ if $i = s$, $P_t^s(x^{2^i}) = 0$ otherwise. The operations P_t^s are derivations on the unstable $\mathcal{A}$-algebra $\Phi^s K$ for any unstable $\mathcal{A}$-algebra K. One can also consider the operations Q_i dual to the elements τ_i. They are useful, when considering classes, on which the Bockstein homomorphism acts non-trivially. One has $Q_i(t) = x^{p^i}$, and $Q_i(x) = 0$, moreover the operations Q_i are derivations.

2. Algebraic Brown-Gitler technology

In this chapter we describe the basic injective objects in the category $\mathcal{U}$. There are three basic types of injective objects.

(i) The Brown-Gitler modules (Section 2.3) which are 'obvious' cogenerators for $\mathcal{U}$.

Recall that $\{C_\alpha\}$, $\alpha \in A$, is a set of cogenerators if any object embeds in a product of C_α's.

(ii) The Carlsson modules (Section 2.6) which are 'obvious' cogenerators for 'reduced' objects in $\mathcal{U}$.

(iii) The tensor product of a Brown-Gitler module by a Carlsson module (Section 2.8).

2.1. Generalities

An object I in an abelian category $\mathcal{C}$ is injective if and only if the functor $M \mapsto \mathrm{Hom}_{\mathcal{C}}(M, I)$ from $\mathcal{C}$ into the category of abelian groups is exact.

In other words, one asks that for any diagram

$$0 \longrightarrow M \overset{j}{\longrightarrow} N$$
$$f \downarrow$$
$$I$$

in $\mathcal{C}$, where the row is exact, there exists $\widetilde{f} : N \to I$ such that $\widetilde{f} \circ i = f$.

35

Example. In the category of abelian groups, divisible groups are injective.

Here are a few basic facts about injective modules in the category $\mathcal{U}$. Most of them are specializations of facts about injective objects in a general abelian category. We refer to [**Br**] to [**Gb**] or to [**Po**] for details.

For convenience, we shall often abbreviate 'injective unstable $\mathcal{A}$-module' to '$\mathcal{U}$-injective'.

Lemma 2.1.1. *An unstable module I is injective if and only if, for any n and any sub-$\mathcal{A}$-module M of $F(n)$, $M \subset F(n)$, any map $M \to I$ extends to a map $F(n) \to I$.*

The proof is exactly like the classical one in [**Gd**] or in [**CE**].

The following statement is true in any locally noetherian abelian category (we follow here Gabriel and assume as part of the definition that a locally noetherian has exact filtered colimits).

Proposition 2.1.2. *Any filtered colimit (in particular any direct sum) of $\mathcal{U}$-injectives is $\mathcal{U}$-injective.*

More generally any filtered colimit of $\mathcal{U}$-injectives is again $\mathcal{U}$-injective. The statement is true in any reasonable abelian category (basically having exact filtered colimit). The converse is also true, if any filtered colimit of injective objects is injective the category is locally noetherian (see sections 4 and 5 in [**Gb**], or [**Po**] Theorem 5.8.7).

An $\mathcal{A}$-module which is finite dimensional as an $\mathbb{F}_p$-vector space in every degree will be said to be of *finite type*.

Proposition 2.1.3 [LZ1]. *Any filtered limit of $\mathcal{U}$-injectives of finite type is $\mathcal{U}$-injective.*

Consider for example an inverse system $\{I_m, \alpha_m : I_m \to I_{m-1}; m \in \mathbb{N}\}$. Let M be a submodule of $F(n)$ and $f = \{f_m\}$, $m \in \mathbb{N}$ be a map from M to $\lim_{m \in \mathbb{N}} I_m$. For each m, choose an extension $\widetilde{f}_m$ of f_m to $F(n)$. As I_0^n is finite, the set $\{\alpha_1 \circ \ldots \circ \alpha_m \circ \widetilde{f}_m\}$, $m \in \mathbb{N}$, is finite. Hence, there is an infinite subset $E_0 \subset \mathbb{N}$ such that $m \in E_0 \Rightarrow \alpha_1 \circ \ldots \circ \alpha_m \circ \widetilde{f}_m = \widetilde{f}'_0$ for some $\widetilde{f}'_0$. We can, for any

m, replace $\widetilde{f}_m$ by $\alpha_{m+1} \circ \ldots \circ \alpha_{m'} \circ \widetilde{f}_{m'}$ for some $m' \in E_0$, $m' > m$. Consequently, one can assume that $\alpha_1 \circ \ldots \circ \alpha_m \circ \widetilde{f}_m = \widetilde{f}_0$ for all m. A classical induction shows that in this way one can construct a compatible sequence of $\widetilde{f}_m$, i.e. a map $\widetilde{f} : F(n) \to \lim_{m \in \mathbb{N}} I_m$ extending f.

2.2. A representability statement

In order to construct $\mathcal{U}$-injectives, we shall proceed as follows. We shall construct contravariant functors from $\mathcal{U}$ to the catégory $\mathcal{E}$ of $\mathbb{F}_p$-vector spaces. These functors will be representable. As they will be left exact, the representing modules will be $\mathcal{U}$-injective. The following lemma says when a contravariant functor $R : \mathcal{U} \to \mathcal{E}$ is representable.

Lemma 2.2.1. *The functor R is representable if and only if it is right exact and transforms direct sums into products.*

We reproduce the proof given in **[LZ1]** and only prove the 'if' part, the other implication being trivial. We need to define an unstable module $B(R)$ and a natural transformation $\gamma : R \to \mathrm{Hom}_{\mathcal{U}}(-, B(R))$. Since we want to have

$$R(F(n)) \cong \mathrm{Hom}_{\mathcal{U}}(F(n)), B(R))$$

we define the unstable module $B(R)$ as a graded $\mathbb{F}_p$-vector space by $B(R)^n = R(F(n))$. The $\mathcal{A}$-module structure is defined as follows: the operation $\theta \in \mathcal{A}$ acts as $R(u_\theta)$, where $u_\theta : F(n + |\theta|) \to F(n)$ is the $\mathcal{A}$-linear map associated to $\theta i_n \in F(n)^{n+|\theta|}$. One checks that this defines an unstable $\mathcal{A}$-module structure on $B(R)$.

The natural transformation is defined as follows. For $M \in \mathcal{U}$ and $x \in R(M)$, let $\gamma_M(x)$ be the ($\mathcal{A}$-linear) map which sends $y \in M$, identified with a map $\check{y} : F(|y|) \to M$, to $R(\check{y})(x) \in R(F(|y|)) = B(R)^{|y|}$. By the very construction $\gamma_{F(n)}$ is an isomorphism. Given $M \in \mathcal{U}$, one looks at the beginning of a free resolution for $M : L_1 \to L_0 \to M \to 0$, where the L_i ($i = 0, 1$) are direct sums of $F(n)$'s. Consider the diagram

$$R(M) \longleftrightarrow R(L_0) \longrightarrow R(L_1)$$

$$\gamma_M \downarrow \qquad\qquad \gamma_{L_0} \downarrow \qquad\qquad \gamma_{L_1} \downarrow$$

$$\mathrm{Hom}_{\mathcal{U}}(M, B(R)) \longleftrightarrow \mathrm{Hom}_{\mathcal{U}}(L_0, B(R)) \longrightarrow \mathrm{Hom}_{\mathcal{U}}(L_1, B(R))$$

Both rows are exact, the top one because R is right exact. The morphisms γ_{L_i} ($i = 0, 1$) are isomorphisms because R transforms direct sums into products. Hence γ_M is also an isomorphism.

Corollary 2.2.2. *Let $\Theta : \mathcal{U} \to \mathcal{C}$ be a covariant functor with values in an abelian category $\mathcal{C}$, which is right exact and transforms direct sums into direct sums. Then Θ has a right adjoint, i.e. there exists a functor $\widetilde{\Theta} : \mathcal{C} \to \mathcal{U}$ such that $\mathrm{Hom}_{\mathcal{C}}(\Theta M, N)$ is naturally isomorphic to $\mathrm{Hom}_{\mathcal{U}}(M, \widetilde{\Theta}N)$ for any $M \in \mathcal{U}$, $N \in \mathcal{C}$.*

Let N be in $\mathcal{C}$. The functor $M \to \mathrm{Hom}_{\mathcal{C}}(\Theta M, N)$ from $\mathcal{U}$ to $\mathcal{E}$ satisfies the hypotheses of (2.2.1). Thus there exists an object in $\mathcal{U}$, which we call $\widetilde{\Theta}N$, such that $\mathrm{Hom}_{\mathcal{C}}(\Theta M, N) \cong \mathrm{Hom}_{\mathcal{U}}(M, \widetilde{\Theta}N)$ for any M in $\mathcal{U}$.

(2.2.3) Examples. The functors Σ and Φ satisfy the hypotheses of (2.2.2). They have right adjoint functors which are denoted by $\widetilde{\Sigma}$ and by $\widetilde{\Phi}$.

The module $\Sigma\widetilde{\Sigma}M$ comes with a map into M, the adjoint of the identity of $\widetilde{\Sigma}M$. One shows easily that $\Sigma\widetilde{\Sigma}M$ embeds in M as the largest suspension contained in M.

In the sequel, we shall denote the adjoint of $\lambda_M : \Phi M \to M$ by $\widetilde{\lambda}_M : M \to \widetilde{\Phi}M$.

Here is another example: let $\mathcal{O}$ be the forgetful functor from $\mathcal{U}'$ into $\mathcal{U}$ ($p > 2$). Thus $\mathcal{O}$ has a right adjoint $\widetilde{\mathcal{O}}$, it is used in **[LZ1]** and **[Z]**.

(2.2.4) Exercise (see **[LZ1]**). Let M be an unstable $\mathcal{A}$-module, $\Sigma\widetilde{\Sigma}M \to M$ be the counit of the adjunction. Let $R^1\widetilde{\Sigma}$ be the first right derived functor of $\widetilde{\Sigma}$. Show that there is an exact sequence:

$$0 \to \Sigma\widetilde{\Sigma}M \to M \xrightarrow{\widetilde{\lambda}_M} \widetilde{\Phi}M \to \Sigma R^1\widetilde{\Sigma}M \to 0.$$

Hint: apply the functor $\operatorname{Hom}_{\mathcal{U}}(\ ,M)$ to the exact sequence

$$0 \to \Phi F(n) \to F(n) \to \Sigma F(n-1) \to 0.$$

(2.2.5) Exercise (see **[LZ1]**). Let p be 2, M be an unstable $\mathcal{A}$-module. Show that $\Omega_1 M$ is isomorphic to $\Sigma \Phi \widetilde{\Sigma} M$.

Note that, if p is 2, the following two conditions are equivalent for any unstable $\mathcal{A}$-module M: (i) λ_M is injective, (ii) $\widetilde{\Sigma} M$ is trivial. If p is an odd prime the first condition implies the second one. The reverse implication is false (exercise: find a counterexample).

2.3. Brown-Gitler modules

Consider the following functor:

$$H_n : M \mapsto M^{n*} = \operatorname{Hom}_{\mathcal{E}}(M^n, \mathbb{F}_p).$$

It is obviously right exact and transforms direct sums into products. Therefore it is representable.

Definition 2.3.1. The n-th Brown-Gitler module $J(n)$ is the representing module for the functor H_n. In other words there is a natural isomorphism $H_n(M) \cong \operatorname{Hom}_{\mathcal{U}}(M, J(n))$.

As H_n is left exact, $J(n)$ is injective in the category $\mathcal{U}$.

Remark. These are really 'Spanier-Whitehead dual' of the modules introduced by Brown and Gitler (see **[BG][BC]** and also **[Ml2]**).

Here are a few facts about the unstable $\mathcal{A}$-modules $J(n)$ which are all consequences of (2.3.1):

$J(0)$ is isomorphic to the unstable $\mathcal{A}$-module $\mathbb{F}_p$;

$J(n)^m = F(m)^{n*}$, thus $J(n)$ is of finite type;

$J(n)^m = 0$ if $m > n$;

$J(n)^n = \mathbb{F}_p$, the 'fundamental class' of $H_n(J(n))$ corresponding to the identity of $J(n)$ will be denoted by b_n. Note that the 'restriction to degree n' $\operatorname{Hom}_{\mathcal{U}}(M, J(n)) \to \operatorname{Hom}_{\mathcal{E}}(M^n, J(n)^n)$ identifies with the isomorphism $\operatorname{Hom}_{\mathcal{U}}(M, J(n)) \cong (M^n)^*$.

The very definition of the unstable $\mathcal{A}$-modules $J(n)$ shows there is an injection

$$M \lhook\joinrel\longrightarrow \prod_n \prod_{u \in H_n(M)} J(n) .$$

Therefore the category $\mathcal{U}$ has enough injective objects.

An element $\theta \in \mathcal{A}$ determines an $\mathcal{A}$-map

$$\bullet\theta : J(n + |\theta|) \to J(n)$$

as follows: $\bullet\theta$ is the map corresponding to the linear form $i_n \circ \theta$: $J(n + |\theta|)^n \to \mathbb{F}_p$, where $\iota_n \in F(n)^n$ is the generator. Alternatively $\bullet\theta$ corresponds to the natural transformation $\lambda \mapsto \lambda \circ \theta$, from $H_{n+|\theta|}(M)$ to $H_n(M)$. One checks that $\bullet(\theta\ \theta')$ is equal to $\bullet\theta \circ \bullet\theta'$. This can be said as follows. If $j \in J(n)^k$, $x \in F(k)^{n-h}$ and $\theta \in \mathcal{A}$ is of degree h, one has

$$(2.3.2) \qquad\qquad (\bullet\theta(j))x = j(\theta x) .$$

The class $\iota_n \otimes \iota_n$ in $H_{m+n}(J(m) \otimes J(n))$ determines an $\mathcal{A}$-linear map $\mu_{m,n} : J(m) \otimes J(n) \to J(m+n)$. Remark that in degree $m + n$ the map $\mu_{m,n}$ is the identity of the ring $\mathbb{F}_p$. If $m = 1$, since $J(1) = \Sigma\,\mathbb{F}_p$ (exercise!), one gets an $\mathcal{A}$-linear map $\mu_{1,n} : \Sigma J(n) \to J(n+1)$.

Proposition 2.3.3. (Mahowald's exact sequences). *Suppose that $p = 2$. Then the following is an exact sequence of $\mathcal{A}$-modules*

$$0 \to \Sigma J(n-1) \xrightarrow{\ \mu_{1,n-1}\ } J(n) \xrightarrow{\ \bullet \mathrm{Sq}^{n/2}\ } J(n/2) \to 0 ,$$

where $\bullet\,\mathrm{Sq}^{n/2}$ and $J(n/2)$ are trivial if n is odd.

Proof. One just applies the functor H_n to the exact sequence $0 \to \Phi F(m) \to F(m) \to \Sigma F(m-1) \to 0$ for all $m \geq 0$.

For $p > 2$, one also has exact sequences. Because of the more complicated formula for Φ, one gets:

Proposition 2.3.4. *For p odd one has the following formula and exact sequences:*

$$\Sigma J(n-1) \cong J(n) \quad \text{if}\ \ n \not\equiv 0\ \text{or}\ 2(2p) ,$$

$$\Sigma J(n-1) \longleftrightarrow J(n) \xrightarrow{\bullet P^{n/2p}} J(n/p) \ \text{if } n \equiv 0(2p),$$

$$\Sigma J(n-1) \longleftrightarrow J(n) \xrightarrow{\bullet \beta P^{n-2/2p}} J((n-2/p)+1) \ \text{if } n \equiv 2(2p).$$

The valuation $v(M)$ of an unstable module M is the smallest n such that M^n is non-zero. The preceding sequences allow computation of the valuation of $J(n)$. For an integer n, denote by $\alpha(n)$ the sum of the coefficients of its p-adic expansion, and by $\mu(n)$ the number of non-zero coefficients.

Proposition 2.3.5. *The valuation* $v(J(n))$ *of* $J(n)$ *is* $2\alpha(n) - \mu(n)$ *(which equals* $\alpha(n)$ *if* $p = 2$).

Proof. Induction on n using Mahowald's exact sequences.

As stated (2.3.3) and (2.3.4) appear in **[LZ1]**. In their original 'dual' form they go back to **[BG][Ch][Mh]**.

Examples of $J(n)$ **at** $p = 2$.

$J(2n+1) \cong \Sigma J(2n)$, $J(2) \cong \tilde{H}^*(\mathbb{RP}^2)$, $J(4) = \tilde{H}^*(\mathbb{RP}^4)$, $J(8)$ is not the reduced mod 2 cohomology of a space (see **[L3]**).

(2.3.6) Exercise. Show that the map $\lambda_{J(n)} : \Phi J(n) \to J(n)$ is equal to the composite of e_n and $\bullet \operatorname{Sq}^n$, e_n being the unique non-zero map from $\Phi J(n)$ to $J(2n)$.

2.4. The bigraded module J_*^*

It is convenient, following Miller **[Ml2]**, to consider the $\mathbb{N} \times \mathbb{N}$ bigraded object $J_*^* = \{J_k^\ell\}$ defined by $J_k^\ell = J(k)^\ell$, where the bidegree of $x \in J_k^\ell$ is denoted by $\|x\| = (|x|_1, |x|_2) = (\ell, k)$.

This object is endowed with a very rich structure:

(2.4.1) It is a left unstable $\mathcal{A}$-module, the structure being induced by those of the unstable $\mathcal{A}$-modules $J(k)$. This action of the Steenrod algebra leaves fixed the second degree and increases the first one.

(2.4.2) The maps $\bullet\theta : J(n) \to J(n - |\theta|)$ induce a right $\mathcal{A}$-module structure on J_*^* which leaves fixed the first degree and decreases the second one. This action is unstable in the sense that $\bullet \operatorname{Sq}^i(x)$

(resp. $\bullet \beta^\varepsilon P^i(x)$, $\varepsilon = 0, 1$), better denoted by $x\,\mathrm{Sq}^i$ (resp. $x(\beta^\varepsilon P^i)$, $\varepsilon = 0, 1$), is zero as soon as $2i > |x|_2$ (resp. $2ip + 2\varepsilon > |x|_2$) if $p = 2$ (resp. $p > 2$).

(2.4.3) The left and the right action commute with each other.

(2.4.4) The maps $\mu_{m,n} : J(m) \otimes J(n) \to J(m+n)$ induce a commutative, associative, unital, bigraded algebra structure on J_*^*.

Remark. At odd primes, this means that

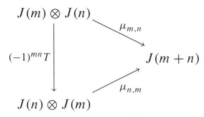

commutes, where $T(a \otimes b) = (-1)^{|a||b|} b \otimes a$.

(2.4.5) There is a Cartan formula for both actions. In the case of the right action, this means that the following diagram commutes (at $p = 2$)

$$
\begin{array}{ccc}
J(m) \otimes J(n) & \xrightarrow{\mu_{m,n}} & J(n+m) \\
\oplus_{k+\ell=i}\,\bullet\mathrm{Sq}^\ell \otimes \bullet\mathrm{Sq}^k \downarrow & & \downarrow \bullet\mathrm{Sq}^i \\
\oplus_{k+\ell=i} J(m-\ell) \otimes J(m-k) & \xrightarrow{\Sigma\mu_{m-\ell,n-k}} & J(n+m-i)
\end{array}
$$

(2.4.6) Denote by Λ the $\mathcal{A}$-linear map (with respect to the left action). Then for $p = 2$

$$\Lambda x = \bullet\mathrm{Sq}^{k/2}(x), \quad J_k^\ell \to J_{k/2}^\ell$$

And for $p > 2$

$$
\begin{aligned}
\Lambda x &= \bullet P^{k/2p}(x), & J_k^\ell &\to J_{k/p}^\ell & \text{if } k \equiv 0\,(2p), \\
\Lambda x &= \bullet\beta P^{(n-2)/2p}(x), & J_k^\ell &\to J_{(k-2)/p}^\ell & \text{if } k \equiv 2\,(2p), \\
\Lambda x &= 0 & & & \text{otherwise}.
\end{aligned}
$$

Then if $p = 2$ one has $\mathrm{Sq}_0\, x = (\Lambda x)^2$. And for $p > 2$ one has

$$P_0\, x = (\Lambda x)^p \quad \text{if } |x| \equiv 0\,(2).$$

Among all these facts, (2.4.1) to (2.4.4) are straightforward; only (2.4.5) and (2.4.6) deserve explanations.

About (2.4.5). The fact that the Cartan formula holds for the left action is classical. This and the definition of $\bullet \operatorname{Sq}^i$ easily show that the Cartan formula also holds for the right action. This implies that the map Λ is multiplicative.

About (2.4.6). It follows from (2.4.5) that Λ is multiplicative. Consider the following diagram ($p = 2$)

$$
\begin{array}{ccc}
\Phi J(k) & \xrightarrow{\;\;\Phi\Lambda\;\;} & \Phi J(k/2) \\
i\downarrow & & i\downarrow \\
\left(J(k)^{\otimes 2}\right)_{\mathfrak{S}_2} & \xrightarrow{\;\;\Lambda\otimes\Lambda\;\;} & \left(J(k/2)^{\otimes 2}\right)_{\mathfrak{S}_2} \\
\widetilde{\mu}_{k,k}\downarrow & & \widetilde{\mu}_{k/2,k/2}\downarrow \\
J(2k) & \xrightarrow{\hspace{3cm}} & J(k)
\end{array}
$$

In this diagram $(M^{\otimes 2})_{\mathfrak{S}_2}$ denotes the coinvariants under the action (by permutation of the factors) of $\mathfrak{S}_2$. More generally, $(M^{\otimes n})_{\mathfrak{S}_n}$ is the quotient of $M^{\otimes n}$ by the submodule generated by elements of the form $\sigma m - m$, for all m in M and all σ in $\mathfrak{S}_n$. The map $\mu_{k,k}$ factors through the coinvariants. This defines $\widetilde{\mu}_{k,k}$, while i is defined by $i\Phi(y) = y \otimes y$ (exercise: check that this is $\mathcal{A}$-linear). We claim that the composite map $\widetilde{\mu}_{k/2,k/2} \circ i \circ \Phi\Lambda$ equals $\lambda_{J(k)}$. To check this, one evaluates both maps on an element y of $(\Phi J(k))^k$ and compares the results going back to the definitions. Then the result follows from the relation $i \circ \Phi\Lambda = (\Lambda \otimes \Lambda) \circ i$. Note that the lower part of the diagram also commutes.

For the case of an odd prime one introduces a subfunctor Φ^e of Φ which is defined as follows: $(\Phi^e M)^n = M^{n/p}$ if $n \equiv 0$ $(2p)$, 0 otherwise. One checks easily that $\Phi^e M$ is a sub-$\mathcal{A}$-module of ΦM for the $\mathcal{A}$-module structure given in (2.2.2). Then, in the preceding proof, one replaces 2 by p and $\Phi J(k)$ by $\Phi^e J(k)$ and everything works the same way.

Here is a theorem which describes the structure of J_*^*.

Theorem 2.4.7 [Ml2]. *If $p = 2$ the bigraded algebra J_*^* is isomorphic to the polynomial (free, commutative) algebra $\mathbb{F}_2[x_i, i \geq 0]$ where $\|x_i\| = (1, 2^i)$, and*

(i) *the left A-module structure is determined by* $\mathrm{Sq}^1 x_i = x_{i-1}^2$ *(recall that* $x_{-1} = 0$*) and the Cartan formula, and*

(ii) *the (multiplicative) map* Λ *is determined by* $\Lambda x_i = x_{i-1}$.

Note. Point (ii) determines the right A-module structure.

Theorem 2.4.8 [Ml2]. *If $p > 2$, the bigraded algebra J_*^* is isomorphic to the free commutative (in the bigraded sense) algebra generated by elements e, x_i, $i \geq 0$, t_i, $i \geq 0$, modulo the relation $e^2 = x_0$, where $\|e\| = (1, 1)$, $\|x_i\| = (2, 2p^i)$, $\|t_i\| = (1, 2p^i)$. Moreover*

(i) *the left A-module structure is determined by $P^1 x_i = x_{i-1}^p$ (recall that $x_{-1} = 0$), $P^1 t_i = 0$, $P^1 e = 0$, $\beta t_i = x_i$, $\beta e = 0$ and the Cartan formula.*

(ii) *the (multiplicative) map Λ is determined by $\Lambda x_i = x_{i-1}$, $\Lambda t_i = t_{i-1}$, where $i > 0$, $\Lambda t_0 = t_0 \beta = e$. This, with $t_i \beta = 0$ if $i > 0$ and $x_i \beta = 0$, determines the right A-module structure on J_*^*.*

Proof of (2.4.7). Consider $T_*^* = \mathbb{F}_2[x_i, i \geq 0]$ as abstractly defined in (2.4.7). One observes that the non-zero element ε_i of $J(2^i)^1 = J_{2^i}^1 = \mathbb{F}_2$ satisfies $\Lambda \varepsilon_i = \varepsilon_{i-1}$ ($\varepsilon_{-1} = 0$) by (2.3.3). Therefore, there is a unique map of bigraded algebras $T_*^* \to J_*^*$ which sends x_i to ε_i and commutes with Λ.

Moreover we have a commutative diagram for any n

$$
\begin{array}{ccccccccc}
0 & \longrightarrow & T_{n-1}^* & \xrightarrow{\times x_0} & T_n^* & \xrightarrow{\Lambda} & T_{n/2}^* & \longrightarrow & 0 \\
& & \downarrow & & \downarrow & & \downarrow & & \\
0 & \longrightarrow & J_{n-1}^* & \xrightarrow{\times \varepsilon_0} & J_n^* & \xrightarrow{\Lambda} & J_{n/2}^* & \longrightarrow & 0
\end{array}
$$

with the usual convention concerning $J_{n/2}^*$ and $T_{n/2}^*$. The first row is exact by elementary algebra. The second row is exact because of (2.3.3). Note that $\mu_{n-1,1}$ is identified with the multiplication by ε_0. The commutativity follows. Now (2.4.7) follows by induction on n.

Proof of 2.4.8. One considers the algebra T_*^* defined by the theorem. Then one produces elements $\varepsilon \in J_1^1$, $\xi_i \in J_{2p^i}^2$, $\tau_i \in J_{2p^i}^1$ such that $\varepsilon^2 = \xi_0$, $\Lambda \xi_i = \xi_{i-1}$ ($\xi_{-1} = 0$), $\Lambda \tau_i = \tau_{i-1}$ ($i > 0$), $\Lambda \tau_0 = \varepsilon$, $\Lambda \varepsilon = 0$ using (2.3.4). These elements determine a map $T_*^* \to J_*^*$, and using the diagrams as above, one gets the result.

2.5. The relation between J_*^* and Milnor's algebra $\mathcal{A}_*$

Recall the structure of the dual algebra $\mathcal{A}_*$ of $\mathcal{A}$ (1.10).

Theorem. *The algebra $\mathcal{A}_*$ is*

(i) *polynomial on generators ξ_i, $i \geq 1$, of degree $2^i - 1$ if $p = 2$;*

(ii) *the tensor product of a polynomial algebra on generators ξ_i, $i \geq 1$, of degree $2(p^i - 1)$ by an exterior algebra on generators τ_i, $i \geq 0$, of degree $2p^i - 1$ if $p > 2$.*

These results suggest that there is a link between $\mathcal{A}_*$ and J_*^*. This has been analyzed in **[BC]** in the context of the original Brown-Gitler modules and in **[Ml2]** in the present context. The link is as follows. The free unstable $\mathcal{A}$-module $F(n)$ is a quotient of $\Sigma^n \mathcal{A}$ (1.6.2). Therefore, by dualization, $J(k)$ should be thought of as a subobject of $\mathcal{A}_*$, specifically $J(k)^{k-n}$ is sent to $\mathcal{A}_*^n$. The map Φ from $\mathcal{A} \otimes \mathcal{A}_*$ to $\mathcal{A}_*$ defined by the formula

$$\Phi(\theta \otimes u) = \Sigma \ u''(\theta) \, u' \, ,$$

where $\delta u = \Sigma u' \otimes u''$, determines a left $\mathcal{A}$-module structure on $\mathcal{A}_*$. This action decreases the degree. A routine check shows that:

(i) the maps $J(k) \to \mathcal{A}_*$ are $\mathcal{A}$-linear, and define, in the obvious way, a map $m : J_*^* \to \mathcal{A}_*$;

(ii) the map $m : J_*^* \to \mathcal{A}_*$ is multiplicative.

We stress that these maps do not preserve degree. By the very definition of the elements ξ_i and τ_i, one gets

Proposition 2.5.1. *If $p = 2$, $m(x_i) = \xi_i$ ($\xi_0 = 1$); if $p > 2$, $m(x_i) = \xi_i$, $m(\tau_i) = t_i$ and $m(e) = 1$.*

Let us prove for example that $m(x_i) = \xi_i$ if $p = 2$. The element ξ_i is determined by the following property: if $u \in H^1 \mathbb{Z}/2$ is the generator, for any $\theta \in \mathcal{A}^{2^i-1}$ one has $\theta u = \langle \xi_i, \theta \rangle u^{2^i}$. Therefore, it is enough to check that $\theta u = \langle m(x_i), \theta \rangle u^{2^i}$. We are done since $\langle m(x_i), \theta \rangle$ is equal to $\langle x_i, \theta i_1 \rangle$, $i_1 \in F(1)^1$ being the generator which can be identified with u.

The following is now straightforward:

Corollary 2.5.2.

(i) *The algebra* $\mathcal{A}_*$ *is the quotient of* J_*^* *by the ideal generated by* $1 - x_0$ *(resp.* $1 - e$ *) if* $p = 2$ *(resp.* $p > 2$ *);*

(ii) *the image of* $J(2k)$ *in* $\mathcal{A}_*$ *under m is the span of the monomials* $\xi_1^{\alpha_1} \ldots \xi_t^{\alpha_t}$ *such that* $2\alpha_1 + 4\alpha_2 + \ldots + 2^t \alpha_t \leq 2k$ *if* $p = 2$ *;*

(iii) *the image of* $J(2k)$ *in* $\mathcal{A}_*$ *under m is the span of the monomials* $\tau_0^{\varepsilon_0} \ldots \tau_t^{\varepsilon_t} \xi_1^{\alpha_1} \ldots \xi_t^{\alpha_t}$ *such that*

$$2\varepsilon_0 + 2\varepsilon_1 + \ldots + 2p\alpha_1 + \ldots + 2p^t \alpha_t \leq 2k \quad \text{if} \quad p > 2.$$

2.6. Carlsson's modules $K(i)$ and reduced unstable $\mathcal{A}$-modules

We now introduce certain unstable $\mathcal{A}$-modules constructed by Carlsson (in the case $p = 2$) in **[Cl1]**. Let us first suppose that $p = 2$ and i is a given integer. The functor

$$M \mapsto \operatorname{Hom}_{\mathcal{E}}(\operatorname{colim}\{M^{2^q i}, \operatorname{Sq}^{2^q i}; q \in \mathbb{N}\}; \mathbb{F}_2)$$

from $\mathcal{U}$ to $\mathcal{E}$ satisfies the hypotheses of Lemma 2.2.1, and it is representable. It is represented by the limit, denoted $K(i)$, of the system

$$\{J(2^q i), \bullet \operatorname{Sq}^{2^{q-1} i}; q \in \mathbb{N}\}.$$

Let $p > 2$ and let $2i$ be a given even integer. The functor

$$M \mapsto \operatorname{Hom}_{\mathcal{E}}(\operatorname{colim}\{M^{2p^q i}, P^{p^q i}; q \in \mathbb{N}\}; \mathbb{F}_p)$$

satisfies the hypotheses of Lemma 2.2.1, and it is represented by the limit, denoted $K(2i)$, of the system

$$\{J(2p^q i), \bullet P^{p^{q-1} i}; q \in \mathbb{N}\}.$$

For $p > 2$ and an odd integer $2i + 1$, one can introduce the functor

$$M \mapsto \operatorname{Hom}_{\mathcal{E}}(\operatorname{colim}\{M^{2(pi+1)p^q}, P^{(pi+1)p^q}; q \in \mathbb{N}\}; \mathbb{F}_p)$$

Note that this colimit is the same as the following one

$$M^{2i+1} \xrightarrow{\beta P^i} M^{2(pi+1)} \xrightarrow{P^{pi+1}} M^{2(pi+1)p} \rightarrow \ldots.$$

This functor is represented by the limit, which we shall denote by $K(2i + 1)$, of the system $\{J(2(pi + 1)p^q, \bullet P^{(pi+1)p^{q-1}}; q \in \mathbb{N}\}$. Note that $K(2i + 1) \cong K(2pi + 2)$.

It follows from (2.3.3) and (2.3.4) that $K(i)^d \cong J(2p^q i)^d$ as soon as q is sufficiently large. Therefore, $K(i)$ is of finite type. All the functors described above are left exact. Thus

Proposition 2.6.1 [LZ1]. *The unstable $\mathcal{A}$-modules $K(i)$ are injective.*

This follows also from (2.1.3).

It is time to introduce two definitions. An unstable $\mathcal{A}$-module is **nilpotent** if the following holds:

for any $x \in M$, $\mathrm{Sq}_0^N x = 0$ as soon as N is sufficiently large, for $p = 2$;

for any $x \in M$, $P_0^N x = 0$ as soon as N is sufficiently large, for $p > 2$ (it is enough to require this for elements of even degree).

For example, a suspension is a nilpotent module. We shall come back to this later.

An unstable $\mathcal{A}$-module M is **reduced** if it does not contain a non-trivial nilpotent module, or equivalently a non-trivial suspension, or equivalently if $\widetilde{\Sigma} M = 0$. If $p = 2$, this is equivalent to saying that Sq_0 is injective.

Proposition 2.6.2 [LZ1]. *The unstable $\mathcal{A}$-modules $K(i)$ are reduced.*

Observe that, by the very definition of Φ and the $K(i)$'s, there is an isomorphism

$$\mathrm{Hom}_{\mathcal{U}}(M, K(i)) = \mathrm{Hom}_{\mathcal{U}}(\Phi M, K(i))$$

for any M. Let $\Sigma M'$ be a suspension included in $K(i)$. As $\lambda_{\Sigma M'}$ is trivial, the map $\mathrm{Hom}_{\mathcal{U}}(\Sigma M', K(i)) \to \mathrm{Hom}_{\mathcal{U}}(\Phi \Sigma M', K(i))$ is also trivial. Therefore $\mathrm{Hom}_{\mathcal{U}}(\Sigma M', K(i)) = 0$ and $\Sigma M' = 0$. This argument shows that

$$\widetilde{\lambda} : K(i) \to \widetilde{\Phi} K(i)$$

is an isomorphism, which is stronger than being reduced (see 2.2.4).

Proposition 2.6.3. *A reduced unstable $\mathcal{A}$-module M embeds in a product of the unstable $\mathcal{A}$-modules $K(i)$.*

This is because the obvious map

$$M \longrightarrow \prod_{i \in \mathbb{N}} \prod_{u \in \mathrm{Hom}_{\mathcal{U}}(M, K(i))} K(i)$$

is injective if and only if M is reduced. It is clear that if this map is injective, M cannot contain a non-trivial suspension because a product of $K(i)$ does not contain a non-trivial suspension. For the other implication, assume first that $p = 2$. In this case, M is reduced if and only if Sq_0 is injective. When this holds any non-zero $x \in M^i$ has a non-trivial image x' in $\mathrm{colim}\{M^{2^q i}, Sq^{2^q i}; q \in \mathbb{N}\}$, and a linear form which is non-trivial on x' corresponds to an $\mathcal{A}$-linear map $M \to K(i)$ which is non-trivial on x. If $p > 2$, we will need the following

Lemma 2.6.4. *Let M be an unstable $\mathcal{A}$-module. Then the following are equivalent:*

(i) *M is reduced;*

(ii) *for any non-zero x in M, there exists $\theta \in \mathcal{A}$ such that $P_0^n \theta x$ is non-zero for every n.*

Remark. One can assume $|\theta x| \equiv 0$ (2) because $|P_0 \theta x| \equiv 0$ (2) for any x, any θ.

Proof. The implication (ii) $\Rightarrow$ (i) is left to the reader. On the other hand, let $x \in M$ be non-zero. Then $\mathcal{A}x$ is not a suspension; this means that there exists an operation $\theta_1 \in \mathcal{A}$ such that $x_1 = P_0 \theta_1 x$ is non-zero. But, $\mathcal{A}x_1$ also is not a suspension. By induction we construct a sequence of non-zero elements $x_i \in M$, $i \geq 1$ such that $x_i = P_0 \theta_i x_{i-1}$ for some $\theta_i \in \mathcal{A}$ ($x_0 = x$). We claim that the degree of $\theta_i x_{i-1}$ has to be even for all sufficiently large i. One considers the element $x_i = P_0 \theta_i P_0 \theta_{i-1} P_0 \theta_{i-2} \ldots P_0 \theta_1 x$. As P_0 on an odd degree element involves β, the operation $P_0 \theta_i P_0 \theta_{i-1} \ldots P_0 \theta_1$ is a sum of monomials of excess at least the number of those x_k, $k < i$, such that $|\theta_{k+1} x_k| \equiv 1$ (2). As x_i is non-zero for any i, the excess stabilizes for large values of i, this implies the claim.

Now if $|x_i| \equiv 0$ (2) for $i \geq i_0$, one shows easily using (1.7.1*) that $P_0^n x_{i_0}$ is non-zero for every n.

Let us come back to the proof of (2.6.3). One wants to produce a map $M \to K(i)$ which is non-trivial on a given non-zero element x in M. For this, we first find an operation θ such that $|\theta x| \equiv 0$ (2) and such that $P_0^n \theta x$ is non-zero for every n. Then, as in the case $p = 2$, we produce a map $M \to K(|\theta|)$ which is non-trivial on θx and consequently on x.

We note here that it is clear that H^*V is reduced if $p = 2$. It is also easily shown that $H^*\mathbb{Z}/p$ is reduced for any p. But the following needs a proof [DLS][LZ1][Z].

Lemma 2.6.5. *Let p be an odd prime. Then the unstable $\mathcal{A}$-module H^*V is reduced for all V.*

We use (2.6.4). Write $H^*V = \mathbb{F}_p[x_1, \ldots, x_d] \otimes E(t_1, \ldots, t_d)$ where $d = \dim V$. Let z be an element of H^*V, and assume $z \neq 0$. We want to find an operation $\theta \in \mathcal{A}$ such that $\theta z \in \mathbb{F}_p[x_1, \ldots, x_d]$ and $\theta z \neq 0$.

Let $(H^*V)_k$ be the $\mathbb{F}_p$-span of those monomials involving at most k of the classes t_i, $(H^*V)_k$ is a sub $\mathcal{A}$-module and

$$\mathbb{F}_p[x_1, \ldots, x_d] = (H^*V)_0 \subset \ldots \subset (H^*V)_d = H^*V$$

Observe that β sends $(H^*V)_k$ to $(H^*V)_{k-1}$.

Assume that z is an element of $(H^*V)_k$ but not in $(H^*V)_{k-1}$. It suffices to find $\theta \in \mathcal{A}$ such that $\theta z \neq 0$ and $\omega z \in (H^*V)_{k-1}$. Write $z = z_k + \omega$, where $z_k \in (H^*V)_k$ is supposed to be non-zero and $\omega \in (H^*V)_{k-1}$. Consider $P^{(|z|-k)/2}z = P^{(|z|-k)/2}z_k + P^{(|x|-k)/2}\omega$; the class $P^{(|z|-k)/2}z_k$ is obtained from z_k by raising to the power p all occurrences of the elements x_i (thus, $\Pi x_i^{\alpha_i} \Pi t_j^{\varepsilon_j}$ where $\varepsilon_j = 0, 1$ and $\Sigma \varepsilon_j = k$ is sent to $\Pi x_i^{p\alpha_i} \Pi t_j^{\varepsilon_j}$). This implies that $\beta P^{(|x|-k)/2}z_k \in (H^*V)_{k-1}$ is non-zero (exercise!). But as $\beta P^{(|x|-k)/2}\omega$ is in $(H^*V)_{k-2}$, we are done.

2.7. Carlsson's bigraded algebra K_*^*

It is obvious from the definition that $K(i) \cong K(2i)$ if $p = 2$, and that $K(2i) \cong K(2ip)$ if $p > 2$. Because of the isomorphism $K(2i + 1) \cong K(2ip + 2)$, we shall not consider $K(2i + 1)$ any longer.

This suggests extending the definition of the $K(i)$'s to all rationals in $\mathbb{N}[1/2]$ for $p = 2$ and all rationals in $2\mathbb{N}[1/p]$ for $p > 2$ by the following formulas

$$K(i) = K(2^q i) \quad \text{for} \quad 2^q i \in \mathbb{N} \quad \text{if} \quad p = 2;$$
$$K(2i) = K(2ip^q) \quad \text{for} \quad 2ip^q \in \mathbb{N} \quad \text{if} \quad p > 2.$$

At $p = 2$, we can define a $\mathbb{N} \times \mathbb{N}[1/2]$-bigraded object K_*^* by $K_i^j = K(i)^j$.

At $p > 2$, we can define a $\mathbb{N} \times 2\mathbb{N}[1/p]$-bigraded object K_*^* by $K_{2i}^j = K(2i)^j$.

By definition of the $K(i)$'s , K_*^* is the (bigraded) limit of the following $\mathbb{N} \times \mathbb{N}[1/2]$ (resp. $\mathbb{N} \times 2\mathbb{N}[1/p]$) bigraded objects:

— $_qJ_*^*$ with $_qJ_i^j = J(2^q i)^j$ if $2^q i \in \mathbb{N}$, 0 otherwise, if $p = 2$;

— $_q\widetilde{J}_*^*$ with $_q\widetilde{J}_{2i}^j = J(2ip^q)^j$ if $2ip^q \in \mathbb{N}$, 0 otherwise, if $p > 2$.

Note that $_0\widetilde{J}_*^*$ is the subalgebra of J_*^* generated by the classes x_i and t_i, $i \geq 0$. The limit is taken over the maps Λ by interpreting them as maps from $_qJ_*^*$ to $_{q-1}J_*^*$ if $p = 2$ (resp. from $_q\widetilde{J}_*^*$ to $_{q-1}\widetilde{J}_*^*$ if $p > 2$). The bigraded object $_qJ_*^*$ should be thought of as a copy of J_*^* 'shifted multiplicatively down' q-times with respect to the second grading. More precisely, if $p = 2$, one has:

$$_qJ_*^* = \mathbb{F}_2[_qx_i, i \geq -q]$$

and

$$\Lambda(_qx_i) = \begin{cases} _{q-1}x_i & \text{if} \quad i > -q + 1 \\ 0 & \text{otherwise}. \end{cases}$$

At $p > 2$, analogous statements hold except that one has to replace J_*^* by $\widetilde{J}_*^*$.

The 'external' maps $\Lambda : {}_q J_*^* \to {}_{q-1} J_*^*$ being multiplicative, K_*^* is endowed with an algebra structure.

The 'internal' maps $\Lambda : {}_q J_*^* \to {}_q J_*^*$ determine a map $\Lambda : K_*^* \to K_*^*$.

Finally, K_*^* is endowed with a left $\mathcal{A}$-module structure. The following theorem of [**LZ1**], implicit in [**Cl1**], describes the structure of K_*^* :

Theorem 2.7.1. *The Carlsson algebra K_*^* is, at $p = 2$, isomorphic to the polynomial algebra $\mathbb{F}_2[\hat{x}_i, i \in \mathbb{Z}]$ where*

(i) $\|\hat{x}_i\| = (1, 2^i)$,

(ii) $\Lambda\hat{x}_i = \hat{x}_{i-1}$, *and*

(iii) *the relation* $\mathrm{Sq}^1 \hat{x}_i = \hat{x}_{i-1}^2$ *and the Cartan formula determine the $\mathcal{A}$-module structure.*

Theorem 2.7.2. *The Carlsson algebra K_*^* is, at $p > 2$, isomorphic to the tensor product $\mathbb{F}_p[\hat{x}_i, i \in \mathbb{Z}] \otimes E(\hat{t}_i, i \in \mathbb{Z})$ where*

(i) $\|\hat{x}_i\| = (2, 2p^i)$, $\|\hat{t}_i\| = (1, 2p^i)$,

(ii) $\Lambda\hat{x}_i = \hat{x}_{i-1}$, $\Lambda\hat{t}_i = \hat{t}_{i-1}$, *and*

(iii) *the relations* $\beta\hat{t}_i = \hat{x}_i$, $P^1\hat{x}_i = \hat{x}_{i-1}^p$ *and the Cartan formula determine the $\mathcal{A}$-module structure.*

Note that K_*^* is no longer a right $\mathcal{A}$-module.

The proof of these two theorems is easy.

Assume for example that $p = 2$. Then $\mathbb{F}_2[\hat{x}_i]$ comes with an algebra map to K_*^*. Indeed define an algebra map $\mathbb{F}_2[\hat{x}_i] \to {}_q J_*^*$ by sending $\hat{x}_i$ to ${}_q x_i$. These maps yield an algebra map $\mathbb{F}_2[\hat{x}_i] \to K_*^*$. This map is clearly injective. It is surjective because K_i^j is isomorphic to ${}_q J_i^j$ for q large enough. This follows from 2.3.3 (2.3.4 for an odd prime).

2.8. Tensor products of injective unstable $\mathcal{A}$-modules

In this section we analyze the tensor product $K(i) \otimes J(n)$ and prove the following results, due to Lannes and Zarati (see [LZ1]):

Theorem 2.8.1. *The unstable $\mathcal{A}$-module $K(i) \otimes J(n)$ is $\mathcal{U}$-injective for all i and all n.*

Corollary 2.8.2. *Let I and J be two $\mathcal{U}$-injectives such that*
(i) $\tilde{\Sigma} J = 0$ *(i.e. J is a reduced unstable $\mathcal{A}$-module), and*
(ii) *I or J is of finite type;*
Then $I \otimes J$ is injective.

Proof of corollary.   Assume for example that I is of finite type. Then, since J is a direct summand in a product $\prod_\alpha K(i_\alpha)$ (2.6.3), $I \otimes J$ is a direct summand of $I \otimes \prod_\alpha K(i_\alpha)$, which is isomorphic to $\prod_\alpha I \otimes K(i_\alpha)$ (here we use that I is of finite type). Now I is by (2.3) a direct summand of a product $\prod_\beta J(n_\beta)$. Therefore $I \otimes J$ is a direct summand of $\prod_\alpha (\prod_\beta J(n_\beta)) \otimes K(i_\alpha)$, which is isomorphic to $\prod_\alpha \prod_\beta J(n_\beta) \otimes K(i_\alpha)$, as $K(i_\alpha)$ is of finite type. The result follows.
    We prove Theorem 2.8.1 for $p > 2$. The case $p = 2$ is similar. We shall need

Lemma 2.8.3. *There exists an integer $k(i, n)$ depending only on i and n such that, in degrees less than $q + k(i, n)$, the following diagram commutes*

$$
\begin{array}{ccc}
J(2p^{q+1}i) \otimes J(n) & \xrightarrow{\mu_{2p^{q+1}i,n}} & J(2p^{q+1}i + n) \\
{\scriptstyle \bullet P^{p^q i} \otimes \mathrm{Id}} \downarrow & & \downarrow {\scriptstyle \bullet P^{p^q i}} \\
J(2p^q i) \otimes J(n) & \xrightarrow{\mu_{2p^q i,n}} & J(2p^q i + n)
\end{array}
$$

We are going to use Proposition 1.2.2. Let $a \otimes b$ be an element in $J(2p^{q+1}i) \otimes J(n)$. Then, by the 'right' Cartan formula (2.4.5), one has

$$
\bullet P^{p^q i}(ab) = (\bullet P^{p^q i}a)b + \sum_{k=1}^{k=p^q i} (\bullet P^{p^q i - k}a)(\bullet P^k b) .
$$

Here xy is the product of x and y in J_*^*. Now the sum on the right runs only from 1 to $[n/2(p-1)]$. For if $k > [n/2(p-1)]$ the integer $n - 2k(p - 1)$ is negative and $J(n - 2k(p - 1))$ is zero.

Note that the integers h such that p^h divides $p^q i - k$ for some $q \in \mathbb{N}$ and some k such that $1 \le k \le n/2(p - 1)$ are obviously bounded by a constant $k_0(i, n)$ which depends only on i and n.

We claim that $\bullet P^{p^\ell} a$ is zero as soon as $0 \le \ell \le k_0(i, n)$ and $|a| < q + k(i, n)$, for a certain constant $k(i, n)$ depending only on i and n. Then the result follows from Proposition 1.2.2 as $|a| \le |a \otimes b|$.

Now the element $\bullet P^{p^\ell} a$ is in $J(2p^{q+1}i - 2p^\ell(p - 1))$ and will be zero as soon as $|a|$ is strictly less than $v(J(2p^{q+1}i - 2p^\ell(p - 1)))$. It is elementary to show, using (2.3.5), that there exists a constant $k'(i, \ell) \in \mathbb{Z}$ such that, for any q

$$q + k'(i, \ell) < v(J(2p^{q+1}i - 2p^\ell(p - 1))) .$$

Now let $k(i, n)$ be $\inf_{0 \le \ell \le k_0(i,n)} k'(i, \ell)$. The lemma follows.

Consider the representing module $L(i, n)$ for the functor

$$M \mapsto \mathrm{Hom}_{\mathcal{E}}(\mathrm{colim}\{M^{2p^q i + n}, P^{p^q i}; q \in \mathbb{N}\}; \mathbb{F}_p) .$$

Then $L(i, n)$ is $\mathcal{U}$-injective, and is the limit of the system

$$\left\{ J(2p^q i + n), \bullet P^{p^{q-1}i}; q \in \mathbb{N} \right\} .$$

Lemma 2.8.3 shows that there is a map

$$\mu_{i,n} : K(i) \otimes J(n) \to L(i, n) .$$

More precisely, if $x \otimes y \in K(i)^k \otimes J(n)^\ell$, define $\widetilde{\mu}_{i,n}(x \otimes y)$ to be the family

$$\left\{ \mu_{2p^q i,n}(x \otimes y) , \ q \text{ such that } q + k(i, n) > |x| + |y| \right\} .$$

In this definition, one identifies x with its image in $J(2^q i)$. Lemma 2.8.3 shows that this defines an element in the limit of the system

$$\left\{ J(2p^q i + n), \bullet P^{p^{q-1}i}; q > |x| + |y| - k(i, n) \right\} ,$$

which is just $L(i, n)$. Theorem 2.8.1 is implied by

Proposition 2.8.4. *The map $\widetilde{\mu}_{i,n}$ is an isomorphism.*

The proof proceeds by induction over n. The case $n = 0$ is trivial. Assuming $\widetilde{\mu}_{i,k}$ is an isomorphism for $k < n$, the only difficult cases are when $n \equiv 0$ or 2 $(2p)$ (see 2.3.4). Here is the case $n \equiv 0$ $(2p)$. There is a commutative diagram where both rows are exact:

$$0 \to \Sigma J(2p^q i) \otimes J(n-1) \longrightarrow J(2p^q i) \otimes J(n) \to J(2p^q i) \otimes J(n/p) \longrightarrow 0$$

$$\Sigma\mu_{2p^q i,n-1} \qquad\qquad \mu_{2p^q i,n} \qquad\qquad \bullet P^{p^{q-1}i} \otimes \mathrm{Id}$$

$$J(2p^{q-1}i) \otimes J(n/p) \to 0$$

$$\mu_{2p^{q-1}i,n/p}$$

$$0 \longrightarrow \Sigma J(2p^q i + n - 1) \longrightarrow J(2p^q i + n) \longrightarrow J(2p^{q-1}i + n/p) \longrightarrow 0$$

Checking the commutativity is left to the reader.

To allow passage to the limit over q in the preceding diagram, one must check that the maps $\bullet P^{p^{q-1}i+n/2p}$ induce a map of the inverse system

$$\left\{ J(2p^q i + n), \bullet P^{p^{q-1}i} ; q \in \mathbb{N} \right\}$$

to the inverse system

$$\left\{ J(2p^{q-1}i + n/2p), \bullet P^{p^{q-2}i} ; q \in \mathbb{N} \right\}.$$

This is implied by the following formula which is a 'right analogue' of (1.7.1*):

For any unstable right $\mathcal{A}$-module M, and any $x \in M$ such that $|x| \equiv 0$ (2), one has

$$(x P^{jp}) P_0 = (x P_0) P^j,$$

where $x P_0 = x P^{|x|/2p}$. *In our case, this reads* $\bullet P_0(\bullet P^{jp} x) = \bullet P^j(\bullet P_0 x)$. *We insist that* $\bullet P_0$ *is understood with respect to the second grading in* J_*^*.

Therefore, we get at the limit a commutative diagram:

$$0 \longrightarrow \Sigma K(i) \otimes J(n-1) \longrightarrow K(i) \otimes J(n) \longrightarrow K(i) \otimes J(n/p) \longrightarrow 0$$

$$0 \longrightarrow \Sigma L(i, n-1) \longrightarrow L(i, n) \longrightarrow L(i, n/p) \longrightarrow 0$$

where the rows are exact. For exactness of the bottom row, one uses that the unstable $\mathcal{A}$-modules $J(k)$ are of finite type. The induction follows.

2.9. The unstable modules $K(i)$ and binary trees

Here are a few remarks about the $K(i)$'s for $p = 2$. They are due, for the first part, to D. Davis and H. Miller [Da][Ml3]. By definition $K(i)$, $i \in \mathbb{N}[1/2]$, has as $\mathbb{F}_2$-basis in degree n those monomials $\prod_{\ell \in \mathbb{Z}} \hat{x}_\ell^{\alpha_\ell}$ such that $\sum_\ell 2^\ell \alpha_\ell = i$ and $\Sigma \alpha_\ell = n$ (almost all α_ℓ being zero).

Assume for the moment that $i = 1$. Consider planar rooted binary trees (rooted means there is a distinguished node), i.e. graphs where each node is binary and the graph is connected. (see Figure 1).

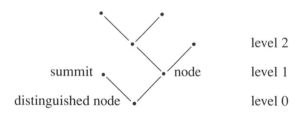

Figure 1

One defines in the obvious way the level of a node or of a summit.

Consider an equivalence relation on such trees. This relation is generated by the following elementary equivalences. At a given level one can move the part of the graph which is above any node onto any summit, for example the two trees of Figure 2 (below) are equivalent. Formally such an object is a sequence of integers $(n_1, \ldots, n_k, \ldots)$, all but a finite number zero, such that:

(1) $0 \le n_1 \le 1$,

(2) $0 \le n_j \le 2n_{j-1}$.

Here n_j is the number of binary nodes at level $j - 1$, and the sum $(\Sigma n_i) + 1$ is the number of summits. Let us call the integer k the height of the (equivalence class of the) tree. Let H_n be the number of equivalence classes of trees with n (binary) nodes.

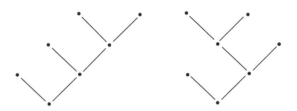

Figure 2

Proposition 2.9.1. *There is a bijection between monomials $\prod_\ell \hat{x}_\ell^{\alpha_\ell}$ such that $\Sigma_\ell 2^\ell \alpha_\ell = 1$, $\Sigma \alpha_\ell = n$, and equivalence classes of binary trees with n summits.*

Proof.   Given a sequence $(n_1, \ldots, n_k)$, one associates to it the monomial $\hat{x}_{-1}^{2n_1-n_2} \hat{x}_{-2}^{2n_2-n_3} \ldots \hat{x}_{-k}^{2n_k}$ if $n_1 = 1$, or x_0 if $n_1 = 0$. One easily checks the required conditions.

On the other hand given a monomial $\prod_\ell \hat{x}_\ell^{\alpha_\ell}$ such that $\Sigma 2^\ell \alpha_\ell = 1$, let k be the largest integer such that α_{-k} is non-zero. If k is zero let $n_1 = n_2 = \ldots = 0$. Otherwise, one observes that $n_i = \alpha_{-i}/2 + \ldots + \alpha_{-k}/2^{k-i+1}$, $1 \le i \le k$, is an integer and that the sequence $(n_1, \ldots, n_k, 0, \ldots)$ satisfies the conditions stated above.

The series $H(t) = \Sigma H_n t^n$ has been computed by G. Andrews. The preceding remarks show that the Poincaré series $P_{K(1)}(t)$ of $K(1)$ is equal to $t H(t)$.

Let

$$a(t) = t + \sum_{j \ge 2}(-1)^{j+1} \frac{t^{2^{j+1}-2-j}}{(1-t)\ldots(1-t^{2^{j-1}-1})}$$

$$b(t) = t + \sum_{j \geq 1} (-1)^j \frac{t^{2^{j+1}-2-j}}{(1-t)\ldots(1-t^{2^j-1})} .$$

Then

Theorem 2.9.2 [Aw]. *The series $H(t)$ is given by the formula*

$$H(t) = 1 + (a(t)/b(t)) .$$

There are several remarkable things about this formula. The first one is that the Poincaré series of the 'Steinberg summand' $L(j)$ of $B(\mathbb{Z}/2)^j$, (see **[MtP]** and Chapter 4), is just

$$P_{L(j)}(t) = \frac{t^{2^{j+1}-2-j}}{(1-t)\ldots(1-t^{2^j-1})} ;$$

hence $b(t) = \sum_{j \geq 1}(-1)^j P_{L(j)}(t)$. This fact deserves an explanation. Secondly, the number H_n satisfies the asymptotic estimates $H_n \sim K\theta^n$ where $K = 0.254505\ldots$, $\theta = 1.794147\ldots$ and θ^{-1} is the smallest positive root of a certain transcendental equation (see **[FP]**). This number θ also appears when one studies the growth of the Λ-algebra (see **[T1]**).

We outline Andrews' proof because facts interesting to the algebraic topologist appear in it (note that there is a misprint in the formula in Andrews' paper). By definition

$$H(t) = \sum_{(n_1,\ldots,n_k,0,\ldots)} t^{n_1+\ldots+n_k} ,$$

$(n_1, \ldots, n_k, 0, \ldots)$ running through the set of sequences subject to the above conditions. Let j be a fixed integer and define $\mu_0(t) = 1$ and $\mu_j(t)$ by

$$\mu_j(t) = \sum_{(n_1,\ldots,n_j,0,\ldots),n_j \neq 0} t^{n_1+\ldots+n_j} ;$$

by definition $H(t) = \sum_j \mu_j(t)$.

It is a fact that $\mu_j(t)$ is the Poincaré polynomial of $J(2^j - 1)$ (exercise 2.9.5 below).

Andrews proves that

$$\sum_0^j (-1)^k \mu_{j-k}(t)\ell_k(t) = (-1)^j t^{2^j-1}\ell_j(t), \quad j \geq 0.$$

In this formula, $\ell_j(t)$ is equal to $P_{L(j)}(t)$ as defined above. In fact summing over n_j, then over n_{j-1}, ... he gets:

$$\sum_0^h (-1)^k \mu_{j-k}(t)\ell_k(t)$$

$$= (-1)^h \sum_{n_1,\ldots,n_{j-h}} t^{n_1+\ldots+n_{j-h-1}+(2^{h+1}-1)n_{j-h}}\ell_h(t),$$

for $0 \leq h \leq j - 1$. Clearly, the result follows. The interesting point (for the algebraic topologist) in this proof is the appearance of the Poincaré polynomials of certain Brown-Gitler modules.

It is possible to extend part of the above discussion to all $K(i)$'s. There is, at least, an analogue of Proposition 2.9.1. One replaces 'rooted binary tree' by the following: setting $i = 2^{\alpha_1} + \ldots + 2^{\alpha_r}$, $\alpha_1 < \alpha_2 < \ldots < \alpha_r$, one considers families of r (rooted, binary) trees, the 'roots' (distinguished points) being at level $-\alpha_1, -\alpha_2, \ldots,$ $-\alpha_r$. The equivalence relation is generated similarly. To prove it one observes that an equation $\sum_j a_j 2^{-j} = 2^{\alpha_1} + \ldots + 2^{\alpha_r}, a_j \geq 0$ 'splits' (non-uniquely) as a collection of equations $2^t = \sum_u a_{t,u} 2^{-u}$ with $a_{t,u} \geq 0$ and $\sum_t a_{t,u} = a_u, 1 \leq t \leq r$.

Example. $i = 3 = 2^1 + 2^0$

The families of Figure 3 below are equivalent.

Exercises.

(2.9.3) Formulate a definition of these equivalence classes of families of trees in terms of sequences of numbers.

(2.9.4) Show that the Poincaré series of $K(3)$ is given by

$$P_{K(3)}(t) = \frac{\sum_{m\geq 1}(-1)^{m+1}t^{2^m}(1 + t^{2^{m-1}})\ell_{m-1}(t)}{\sum_{m\geq 0}(-1)^m \ell_m(t)}.$$

In this case one looks for sequences of integers $(c_0, \ldots, c_m)$ such that $1 \leq c_0 \leq 3, \ldots, 1 \leq c_i \leq 2c_{i-1}, \ldots, \sum_{c_0} \sum_{c_1} \ldots t^{3+c_0+c_1+\ldots}$ turns out to be $P_{K(3)}(t) - t P_{K(1)}(t)$.

(2.9.5) Show that $\mu_j(t)$ is the Poincaré polynomial of $J(2^j - 1)$.

M. Tangora in **[T2]** has made further computations that involve the case of an odd prime.

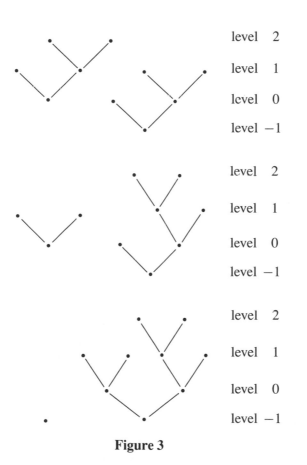

Figure 3

3. $\mathcal{U}$-Injectivity of the mod p cohomology of elementary abelian p-groups and Lannes' functor T_V

In this important chapter we prove the injectivity in the category $\mathcal{U}$ of the unstable modules $H^*V \otimes J(n)$. Then, in Section 3.2, we introduce the functor T_V. In Sections 3.3 and 3.4 we give examples of T_V-computations. In Section 3.5 we prove that T_V commutes with tensor products. Then, in Section 3.8, we show that T_V restricts to a functor on the category $\mathcal{K}$ and give examples in Sections 3.9 and 3.10. In particular we prove the non-linear injectivity of H^*V. Sections 3.11 to 3.14 are concerned with the classification of injective modules in the category $\mathcal{U}$.

3.1. $\mathcal{U}$-Injectivity of $H^*V \otimes J(n)$

In this section we prove the following fundamental theorem

Theorem 3.1.1 [Cl1][Ml2][LZ1]. *Let* V *be an elementary abelian* p-*group. Then the unstable* $\mathcal{A}$-*module* $H^*V \otimes J(n)$ *is* $\mathcal{U}$-*injective for all* n.

Carlsson proved that $\widetilde{H}^*\mathbb{Z}/2$ is a direct factor in $K(1)$ when $p = 2$. Then Miller observed that this implies that $\widetilde{H}^*\mathbb{Z}/2$ is $\mathcal{U}$-injective, and proved an analogous result for $p > 2$. Later, Lannes and Zarati extended the result to all V and all n. The case of any V,

$n = 0$ was implicit in [Cl1] and [Ml2]. Note that it is equivalent to prove the result, for $\widetilde{H}^*V$ or for H^*V, as $H^*V \cong \mathbb{F}_p \oplus \widetilde{H}^*V$ and as $\mathbb{F}_p \cong J(0)$ is $\mathcal{U}$-injective.

Proof. We first show that $\widetilde{H}^*\mathbb{Z}/p$ (respectively $\widetilde{H}^*\mathbb{Z}/2$) is a direct summand in K_*^*.

Assume that p is an odd prime. Following Carlsson and Miller we construct a map $\gamma : \widetilde{H}^*\mathbb{Z}/p \to K_*^*$ as follows. Recall that $\widetilde{H}^*\mathbb{Z}/p$, as an $\mathcal{A}$-module, is the direct sum of submodules H_i, $1 \le i \le p - 1$, defined by $(H_i)^k = \widetilde{H}^k\mathbb{Z}/p$ if $k \equiv 2i$ or $2i - 1$ mod $(p - 1)$, and $(H_i)^k = 0$ otherwise. For each $1 \le i \le p - 1$ let $\gamma_i : H_i \to K_{2i}^* \equiv K(2i)$ be the map corresponding to a given non-zero element of

$$(\text{colim}\{H^{2p^q i}\ \mathbb{Z}/p, \ P^{p^q i}; q \in \mathbb{N}\})^* \cong \mathbb{F}_p .$$

The class tx^{i-1} generates $(H_i)^{2i-1}$; observe that $(H_i)^m = 0$ if $m < 2i - 1$. One checks easily that K_{2i}^{2i-1}, $1 \le i \le p - 1$, is generated by $\hat{t}_0\hat{x}_0^{i-1}$. By definition of γ_i this implies that, up to a non-zero scalar,

$$(3.1.2) \qquad \gamma_i(tx^{i-1}) = \hat{t}_0\hat{x}_0^{i-1}, \quad 1 \le i \le p - 1 .$$

Note that $\gamma_1(tx^p)$ is not equal to $\hat{t}_0\hat{x}_0^p$.

Next we let $g : K_*^* \to H^*\mathbb{Z}/p$ to be the unique algebra map such that $g(\hat{x}_j) = x$ and $g(\hat{t}_j) = t$ (recall that $H^*\mathbb{Z}/p = \mathbb{F}_p[x] \otimes E(t)$).

Lemma 3.1.3. *The map g is $\mathcal{A}$-linear.*

The proof is left to the reader (use 2.7.1 or 2.7.2).

We now show that the map $g \circ \gamma_i : H_i \to H^*\mathbb{Z}/p$ is an isomorphism onto $H_i \subset H^*\mathbb{Z}/p$. For this observe that $g \circ \gamma_i$ is an isomorphism in degree $2i - 1$, and that

Lemma 3.1.4. *An $\mathcal{A}$-linear map $v : H_i \to H_i$ which is an isomorphism in degree $2i - 1$ is an isomorphism.*

We first claim that for any non-trivial element $z \in H_i$, the intersection of $\mathcal{A}z$ and $\mathcal{A}tx^{i-1}$ is non-trivial.

As $\beta tx^{n-1} = x^n$, it is enough to consider the case where $z = x^n$. Let the p-adic expansion of an integer n be $\alpha_h p^h + \alpha_{h+1} p^{h+1} +$

$\ldots + \alpha_t p^t$, where $\alpha_h \neq 0$ and $\alpha_t \neq 0$ for $0 \leq \alpha_i \leq p - 1$. The classes tx^{n-1} and x^n belong to H_i if and only if $\sum_\ell \alpha_\ell \equiv i \ (p)$. The classes tx^{n-1} and x^n belong to $\mathcal{A}tx^{i-1}$ if and only if $\sum_\ell \alpha_\ell = i$. Consider the element $P^{\alpha_h p^h} x^n$. The sum of the coefficients in the p-adic expansion of $|P^{\alpha_h p^h} x^n|/2 = n + \alpha_h p^h (p^h - 1)$ is less than or equal to $\sum_{n \leq \ell \leq t} \alpha_\ell$. Write this expansion as $\beta_{h'} p^{h'} + \ldots + \beta_{t'} p^{t'}$, $0 \leq h' \leq t'$, with $\beta_{h'}$ and $\beta_{t'}$ non-zero. If the equality holds above, and if $t - h > 0$, necessarily $t' - h' < t - h$. Then, by induction, as $P^{\alpha_h p^h} x^n$ is non-zero, one gets a class of the form $x^{ip^N} \in \mathcal{A}z$ for N large enough. The claim follows.

Now let us prove the lemma. If $v(z) = 0$, then $v(x^{ip^N}) = 0$ for N large. This implies that $v(x^i) = 0$ and $v(tx^{i-1}) = 0$, which is a contradiction.

Now $g \circ \gamma_i$ is an isomorphism onto H_i for any $1 \leq i \leq p - 1$. Therefore, H_i is a direct summand of K_{2i}^*, and $\tilde{H}^* \mathbb{Z}/p$ is a direct summand of $\oplus_1^{p-1} K_{2i}^*$. Consequently, $\tilde{H}^* \mathbb{Z}/p$ is $\mathcal{U}$-injective.

Then, as $H^* \mathbb{Z}/p$ is a reduced unstable $\mathcal{A}$-module, Theorem 3.1.1 follows from Corollary 2.8.2 by induction on dim V.

3.2. Lannes' functor T_V

The object of this section is to formulate Theorem 3.1.1 in a functorial way, as has been done by Lannes. Let L be an unstable $\mathcal{A}$-module of finite type.

Theorem 3.2.1 [L1]. *The functor $M \mapsto L \otimes M$ from $\mathcal{U}$ to itself has a left adjoint, which we shall denote by $N \mapsto (N : L)_\mathcal{U}$. In other words, there is a natural isomorphism*

$$\mathrm{Hom}_\mathcal{U}((N : L)_\mathcal{U}, M) \cong \mathrm{Hom}_\mathcal{U}(N, L \otimes M)$$

for all M and all N.

The functor $N \mapsto (N : L)_\mathcal{U}$ should be thought of as a 'division functor'.

The division by $\mathbb{F}_p$ is the identity functor, the division by $\Sigma \mathbb{F}_p$ is the functor Ω.

In the case where $L = H^*V$ Lannes denoted this functor by T_V. Theorem 3.1.1 implies

Theorem 3.2.2 [L1]. *The functor T_V is exact.*

In all the sequel, we shall denote $T_{\mathbb{F}_p}$ by T. Note that $T_V \cong T^{\dim V}$.

Proof of Theorem 3.2.2. Let $0 \to M' \to M \to M'' \to 0$ be an exact sequence. Saying that $0 \to T_V(M') \to T_V(M) \to T_V(M'') \to 0$ is exact in degree n is equivalent to saying that

$$0 \to \mathrm{Hom}_{\mathcal{U}}(M'', H^*V \otimes J(n)) \to \mathrm{Hom}_{\mathcal{U}}(M, H^*V \otimes J(n))$$
$$\to \mathrm{Hom}_{\mathcal{U}}(M', H^*V \otimes J(n)) \to 0$$

is exact. This is implied by the $\mathcal{U}$-injectivity of $H^*V \otimes J(n)$.

As H^*V is isomorphic to $\mathbb{F}_p \oplus \widetilde{H}^*V$, the functor T_V is naturally equivalent to the 'direct sum' of the identity functor and of the division by $\widetilde{H}^*V$. This last functor is denoted by $\overline{T}_V$ (respectively by $\overline{T}$ if $V \cong \mathbb{F}_p$). It is obviously exact.

The existence of T_V is a consequence of Freyd's adjoint functor theorem (see **[Fr]** or **[ML]**). The main point is that the functor $M \mapsto H^*V \otimes M$ commutes with limits as H^*V is of finite type. There is also a 'solution set condition' to check, but this is easy. We reproduce here the proof given in- **[L1]**. This proof is very close to a proof of Freyd's theorem. We are going to describe 'standard' projective resolutions in the category $\mathcal{U}$. Let $\mathcal{O}$ be the forgetful functor from $\mathcal{U}$ in $\mathcal{E}_{\mathrm{gr}}$ (graded $\mathbb{F}_p$-vector spaces). Then $\mathcal{O}$ has a left adjoint $\mathcal{F}$ characterized by

$$\mathrm{Hom}_{\mathcal{U}}(\mathcal{F}(E), M) \cong \mathrm{Hom}_{\mathcal{E}_{\mathrm{gr}}}(E, \mathcal{O}(M))$$

for any $E \in \mathcal{E}_{\mathrm{gr}}$, $M \in \mathcal{U}$. In this formula $\mathcal{F}(E)$ is the unstable module $\oplus_n (E^n \otimes F(n))$, where E^n is considered as an unstable module concentrated in degree zero; $\mathcal{F}(E)$ is the 'free unstable module' generated by E. Let F be $\mathcal{F} \circ \mathcal{O}$. The adjunction formula yields two natural transformations $\eta : F \to \mathrm{Id}$, $\nu : F \to F^2$. Recall that $\eta_M : F(M) \to M$ is the adjoint of the identity $\mathcal{O}(M) \to \mathcal{O}(M)$ and that $\nu_M : F(M) \to F^2(M)$ is $\mathcal{F}$ applied to the morphism which is the adjoint of Id : $F(M) \to F(M)$. The following diagrams commute for any $M \in \mathcal{U}$:

(3.2.3)

$$
\begin{array}{ccc}
F(M) & \xrightarrow{\nu_M} & F^2(M) \\
\downarrow{\scriptstyle \nu_M} & \searrow{\scriptstyle \text{id}} & \downarrow{\scriptstyle F(\eta_M)} \\
F^2(M) & \xrightarrow[\eta_{F(M)}]{} & F(M)
\end{array}
$$

and

(3.2.3*)

$$
\begin{array}{ccc}
F(M) & \xrightarrow{\nu_M} & F^2(M) \\
\downarrow{\scriptstyle \nu_M} & & \downarrow{\scriptstyle \nu_{F(M)}} \\
F^2(M) & \xrightarrow[F(\nu_M)]{} & F^3(M)
\end{array}
$$

The set (F, η, ν) is called a comonad on the category $\mathcal{U}$.

Recall the definition of a simplicial object in a category $\mathcal{C}$ **[M]**. Let Δ be the category whose objects are the sets $[n] = \{0, 1, \ldots, n\}$, $n \geq 0$, and the morphisms $\Delta([n], [m])$ are the monotonic maps $\mu : [n] \to [m]$, that is maps such that $i \leq j \Rightarrow \mu(i) \leq \mu(j)$. A simplicial object in $\mathcal{C}$ is a contravariant functor from Δ to $\mathcal{C}$.

It can be seen also as a graded object, $\{C_n\}_{n \in \mathbb{N}}$ provided with morphisms:

— $d_i : C_n \to C_{n-1}$, $0 \leq i \leq n$,
— $s_j : C_n \to C_{n+1}$, $0 \leq j \leq n$,

satisfying the following identities:

(3.2.4)

$$
\begin{cases}
d_i d_j = d_{j-1} d_i , & i < j , \\
d_i s_j = s_{j-1} d_i , & i < j , \\
d_i s_j = \text{Id} , & i = j, j+1 , \\
d_i s_j = s_j d_{i-1} , & i > j+1 , \\
s_i s_j = s_j s_{i-1} , & i > j .
\end{cases}
$$

Using the comonad defined above, for any M in $\mathcal{U}$, one defines a simplicial object $F_\bullet(M)$ in $\mathcal{U}$ by:

— $F_\bullet(M)_n = F^{n+1}(M)$,
— $d_i : F_\bullet(M)_n \to F_\bullet(M)_{n-1}$ is given by $d_i = F^i(\eta_{F^{n-i}(M)})$, $0 \leq i \leq n$,

— $s_j : F_\bullet(M)_n \to F_\bullet(M)_{n+1}$ is given by $s_j = F^j(\nu_{F^{n-j}(M)})$, $0 \le j \le n$.

It is standard that the simplicial identities hold. Moreover $F_\bullet(M)$ is augmented by the map $\varepsilon = \eta_M : F(M) \to M$ since $\varepsilon \circ d_0 = \varepsilon \circ d_1$.

The map of graded $\mathbb{F}_p$-vector spaces $h_M : \mathcal{O}(M) \to \mathcal{O}(F(M))$ which is adjoint to the identity of $F(M)$ is such that

$$(3.2.5) \qquad\qquad d_i \circ h = h \circ d_{i-1}, \quad i \ge 1,$$

and

$$(3.2.5^*) \qquad\qquad d_0 \circ h = 0.$$

This gives a homotopy to zero of the identity map of the following complex

$$\to F_\bullet(M)_n \xrightarrow{\Sigma(-1)^i d_i} F_\bullet(M)_{n-1} \to \ldots \to F_\bullet(M)_0 \to M \to 0$$

which is therefore a free resolution of M. The interest of this method is that it gives functorial resolutions [**Gd**].

Now let L be an unstable module which is of finite type. Let us show, following Lannes, that the functor $M \mapsto L \otimes M$ of $\mathcal{U}$ to itself has a left adjoint $N \mapsto (N : L)_{\mathcal{U}}$. We start by defining $(N : L)_{\mathcal{U}}$ for N a free unstable A-module, i.e. for N of the form $\mathcal{F}(E)$, $E \in \mathcal{E}_{\mathrm{gr}}$, by the formula

$$(3.2.6) \qquad\qquad (\mathcal{F}(E) : L)_{\mathcal{U}} = \mathcal{F}((E : \mathcal{O}(L))_{\mathcal{E}_{\mathrm{gr}}}).$$

Here, if H is a given graded $\mathbb{F}_p$-vector space of finite type (of finite dimension in each degree), the functor $E \mapsto (E : H)_{\mathcal{E}_{\mathrm{gr}}}$ is the left adjoint of $E \mapsto H \otimes E$; it is given by the formula

$$(E : H)^i_{\mathcal{E}_{\mathrm{gr}}} = \oplus_k E^{i+k} \otimes H^{k*}.$$

The adjunction formulas show that this definition of $(\mathcal{F}(E) : L)_{\mathcal{U}}$ makes sense.

Let $\varphi : \mathcal{F}(E) \to \mathcal{F}(E')$ be any A-linear map. It is necessary to define the induced A-linear map

$$(\varphi : L)_{\mathcal{U}} : (\mathcal{F}(E) : L)_{\mathcal{U}} \to (\mathcal{F}(E') : L)_{\mathcal{U}}.$$

Indeed the induced map is obvious if φ is induced by a graded vector space map, but otherwise it needs to be defined. Let us consider

$\omega_E : (\mathcal{F}(E) : L)_{\mathcal{E}_{\mathrm{gr}}} \to \mathcal{F}(E : L)_{\mathcal{E}_{\mathrm{gr}}} = (\mathcal{F}(E) : L)_{\mathcal{U}}$ the adjoint of the unique $\mathcal{A}$-linear extension of the composite

$$E \to (E : L)_{\mathcal{E}_{\mathrm{gr}}} \otimes L \to \mathcal{F}(E : L)_{\mathcal{E}_{\mathrm{gr}}} \otimes L .$$

The map $(\varphi : L)_{\mathcal{U}}$ is the unique $\mathcal{A}$-linear extension of the composite map

$$(E : L)_{\mathcal{E}_{\mathrm{gr}}} \to (\mathcal{F}(E') : L)_{\mathcal{E}_{\mathrm{gr}}} \to \mathcal{F}(E' : L)_{\mathcal{E}_{\mathrm{gr}}} .$$

Here the right map is $\omega_{E'}$, and the left one is induced by the composite of the inclusion $E \to F(E)$ with φ.

It is easily checked that

$$\omega_{E'} \circ (\varphi : L)_{\mathcal{E}_{\mathrm{gr}}} = (\varphi : L)_{\mathcal{U}} \circ \omega_E ,$$

and that

$$(\varphi : L)_{\mathcal{U}} \circ (\psi : L)_{\mathcal{U}} = (\varphi \circ \psi : L)_{\mathcal{U}} .$$

Moreover, it is clear that if $f \in \mathrm{Hom}_{\mathcal{U}}(\mathcal{F}(E'), L \otimes N)$ is adjoint to $\tilde{f} \in \mathrm{Hom}_{\mathcal{U}}((\mathcal{F}(E') : L)_{\mathcal{U}}, N)$, then $f \circ \varphi$ is adjoint to $\tilde{f} \circ (\varphi : L)_{\mathcal{U}}$. Again, to simplify notation, we have omitted the forgetful functors.

Then considering the standard resolution of N

$$\ldots \to F_\bullet(N)_1 \xrightarrow{d_0 - d_1} F_\bullet(N)_0 \xrightarrow{\varepsilon} N \to 0 ,$$

one defines
(3.2.7)
$$(N : L)_{\mathcal{U}} = \mathrm{Coker}(d_0 - d_1 : L)_{\mathcal{U}} : (F_\bullet(N)_1 : L)_{\mathcal{U}} \to (F_\bullet(N)_0 : L)_{\mathcal{U}} .$$

By construction, this is a well defined covariant functor of N. A routine check shows that it is a left adjoint for $M \mapsto L \otimes M$. Note that $(N : L)_{\mathcal{U}}$ is a contravariant functor in L.

3.3. First examples of T_V-computations

In this section we describe examples a few of T_V-computations. The first ones follow from generalities about adjoint functors. The second one depends also on a property of the suspension functor. The third one depends on the structure of H^*V as an $\mathcal{A}$-module. The last one uses exactness of T.

Lemma 3.3.1. *For any* n, *the unstable* $\mathcal{A}$-*module* $T(F(n))$ *is isomorphic to the direct sum* $\bigoplus_{0 \leq i \leq n} F(i)$.

By definition, $\text{Hom}_{\mathcal{U}}(T(F(n)), N)$ is isomorphic to

$$\text{Hom}_{\mathcal{U}}(F(n), H^*\mathbb{Z}/p \otimes N)$$

which is isomorphic to $\bigoplus_{i=0,\ldots,n} N^i$. The result follows. In fact, (3.3.1) is implicit in the construction of the general division functors, and depends only on the structure of $H^*\mathbb{Z}/p$ as graded $\mathbb{F}_p$-vector space.

Proposition 3.3.2. *The functor* T_V *commutes with colimits.*

This holds for any left adjoint.

Next we study the behaviour of T_V with respect to the functors Ω and Σ.

Proposition 3.3.3. *The natural map* $T_V \Omega M \to \Omega T_V M$ *is an isomorphism for any unstable* $\mathcal{A}$-*module* M.

We have a sequence of adjunction isomorphisms:

$\text{Hom}_{\mathcal{U}}(\Omega T_V M, N)$

$$\cong \text{Hom}_{\mathcal{U}}(T_V M, \Sigma N) \cong \text{Hom}_{\mathcal{U}}(M, H^*V \otimes \Sigma N)$$

$$\cong \text{Hom}_{\mathcal{U}}(\Omega M, H^*V \otimes N) \cong \text{Hom}_{\mathcal{U}}(T_V \Omega M, N).$$

Only the third isomorphism in this sequence is not completely formal. It depends on the isomorphism $H^*V \otimes \Sigma N \cong \Sigma(H^*V \otimes N)$. Taking $N = \Omega T_V M$ the image of the identity map under these isomorphisms is the desired isomorphism.

Consider the case of the suspension. Suspending the adjoint of the identity of $T_V M$, one gets a natural map $\Sigma M \to H^*V \otimes \Sigma T_V M$. The adjunction yields a natural map $T_V \Sigma M \to \Sigma T_V M$.

Proposition 3.3.4. *The natural map* $T_V \Sigma M \to \Sigma T_V M$ *is an isomorphism for any unstable* $\mathcal{A}$-*module* M.

Recall the right adjoint $\widetilde{\Sigma}$ of Σ (3.2.3). The proof of (3.3.4) looks the same one of (3.3.3) as soon as one has the following lemma (compare with [**LZ1**]):

Lemma 3.3.5. *For any reduced unstable $\mathcal{A}$-module M and any unstable $\mathcal{A}$-module N, the natural map*

$$M \otimes \widetilde{\Sigma} N \to \widetilde{\Sigma}(M \otimes N)$$

is an isomorphism.

Let us prove Lemma 3.3.5. It is clear that $M \otimes \Sigma \widetilde{\Sigma} N$ is contained in $\Sigma \widetilde{\Sigma}(M \otimes N)$. Thus, one has to prove that an element z of degree $n+1$ in $M \otimes N$ is the image of $\Sigma \iota_n \in (\Sigma F(n))^{n+1}$ by an $\mathcal{A}$-linear map φ, if and only if z belongs to $M \otimes \Sigma \widetilde{\Sigma} N \subset M \otimes N$.

Assume that $p = 2$. Write $\varphi(\Sigma \iota_n)$ as $\sum m_i \otimes n_i$. Assume that the m_i are linearly independent. As $\mathrm{Sq}_0 \, \varphi(\Sigma \iota_n) = 0$, the Cartan formula gives $\sum \mathrm{Sq}_0 \, m_i \otimes \mathrm{Sq}_0 \, n_i = 0$. As Sq_0 is injective on M the $\mathrm{Sq}_0 \, m_i$ are linearly independent. Hence the $\mathrm{Sq}_0 \, n_i = 0$, $0 \le i \le n+1$, and n_i belongs to $\Sigma \widetilde{\Sigma} N$ for all i.

If p is odd, we use Lemma 2.6.4 and we proceed as follows. Write $\varphi(\Sigma \iota_n)$ as $\sum_i m_i \otimes n_i$ with $|m_i| + |n_i| = n + 1$ for all i. If the elements m_i belong for all i to the largest evenly graded sub-$\mathcal{A}$-module of M the proof works nearly exactly as above. One checks that $P_0 \theta n_i = 0$ for all $\theta \in \mathcal{A}$ and all i by increasing induction on $|\theta|$ and decreasing induction on $|n_i|$.

Otherwise suppose that $\varphi(\Sigma \iota_n)$ does not belong to $M \otimes \Sigma \widetilde{\Sigma} N$. If, at least one element m_i does not belong to the largest evenly graded sub-$\mathcal{A}$-module of M one observes easily, using Lemma 2.6.4, that there exists an operation P^I, involving β, such that $P^I \varphi(\Sigma \iota_n)$ is non-zero and, again, does not belong to $M \otimes \Sigma \widetilde{\Sigma} N$. Such a process can be done at most n times. Indeed, an operation involving β at least $n + 1$ times is trivial on $\varphi(\Sigma \iota_n)$. Thus, after a finite number of steps all the elements on the right belong to the largest evenly graded sub $\mathcal{A}$-module of M. In that case the same argument as above leads to a contradiction.

We have used the following relations, which are direct consequences of the Cartan formula and of the definitions given in (1.7.1) and (1.7.1*):

For any x (resp. y) in an unstable $\mathcal{A}$-module M (resp. N), one has

$$\mathrm{Sq}_0(x \otimes y) = \mathrm{Sq}_0 x \otimes \mathrm{Sq}_0 y \quad \text{if } p = 2,$$
$$P_0(x \otimes y) = P_0 x \otimes P_0 y \quad \text{if } p > 2, \text{ and } |x| \text{ or } |y| \equiv 0 \ (2).$$

An unstable $\mathcal{A}$-module M is said to be **locally finite** if, for any x in M, the span of x over $\mathcal{A}$ is finite, in other words for any x in M only a finite number of elements of $\mathcal{A}$ act non-trivially on x.

Proposition 3.3.6. *Let M be a locally finite unstable $\mathcal{A}$-module. Then $T_V(M)$ is isomorphic to M.*

Proof. By Lemma 3.3.1, we know that $T_V(\mathbb{F}_p)$ is isomorphic to $\mathbb{F}_p$. By Proposition 3.3.3, we conclude that $T_V(\Sigma^n \mathbb{F}_p) \cong \Sigma^n \mathbb{F}_p$, for any $n \geq 0$. Let M be an unstable $\mathcal{A}$-module which is trivial in large degrees. By filtering M by degree and using the exactness of T_V, one gets $T_V(M) \cong M$. The general case follows using a colimit argument (Proposition 3.3.2). A converse to this proposition will be proved later.

3.4. Commutation of T_V and Φ

In this section we define a natural map $T_V \Phi M \to \Phi T_V M$ and show that this is an isomorphism.

We need a natural map $\Phi(M \otimes N) \to \Phi M \otimes \Phi N$. If $p = 2$, it is the obvious isomorphism between $\Phi(M \otimes N)$ and $\Phi M \otimes \Phi N$. If p is odd, these two modules are not, in general, isomorphic and one has to give a definition.

Let L, M and N be unstable $\mathcal{A}$-modules, $\varepsilon : L \to M \otimes N$ be an $\mathcal{A}$-linear map. Then the following formulas define an $\mathcal{A}$-linear map $\Phi(\varepsilon) : \Phi L \to \Phi M \otimes \Phi N$.

(3.4.1). If $x \in L$, $|x| \equiv 0\,(2)$ and $\varepsilon(x) = \sum_i m_i \otimes n_i + \sum_j m_j' \otimes n_j'$ with $|m_i| \equiv |n_i| \equiv 0$ (2) and $|m_j'| \equiv |n_j'| \equiv 1$ (2), then

$$\Phi(\varepsilon)(\Phi x) = \sum_i \Phi m_i \otimes \Phi n_i \,.$$

If $x \in L$, $|x| \equiv 1$ (2) and if $\varepsilon(x) = \sum_i m_i \otimes n_i$, then

$$\Phi(\varepsilon)(\Phi x) = \sum_i \Phi m_i \otimes \Phi n_i \,.$$

A routine check shows that $\Phi(\varepsilon)$ is indeed well defined and $\mathcal{A}$-linear.

The natural map $T_V \Phi M \to \Phi T_V M$ is defined as follows. Consider the adjoint of the identity of $T_V M$

$$M \to H^* V \otimes T_V M \,;$$

apply (3.4.1) and compose with

$$\lambda_{H^*V} \otimes \mathrm{Id} : \Phi H^* V \otimes \Phi T_V M \to H^* V \otimes \Phi T_V M$$

to obtain a map

$$\Phi M \to H^* V \otimes \Phi T_V M \,;$$

adjointing again, yields the required map.

In the following lemma we assume that the prime is 2.

Lemma 3.4.2. *The map* $T_V(\lambda_M) : T_V \Phi M \to T_V M$ *is the composite of the natural map* $T_V \Phi M \to \Phi T_V M$ *defined above with the map* $\lambda_{T_V M} : \Phi T_V M \to T_V M$.

Proof. Using the definition of the maps involved, one reduces to showing that the following diagram is commutative

$$
\begin{array}{ccc}
\Phi M & \xrightarrow{\quad\lambda_M\quad} & M \\
\downarrow & & \downarrow \\
\Phi H^* V \otimes \Phi T_V M & \xrightarrow{\lambda_H^* V \otimes \lambda_{TV} M} & H^* V \otimes T_V M
\end{array}
$$

which is done using the formulas 3.4.1.

Proposition 3.4.3. *The natural map* $T_V \Phi M \to \Phi T_V M$ *is an isomorphism for any unstable* $\mathcal{A}$-*module* M.

As Φ is right exact (being a left adjoint), it is enough to check the proposition for the unstable $\mathcal{A}$-modules $F(n)$, $n \geq 0$. But Lemma 3.4.2, Propositions 3.3.3 and 3.3.4 show that there is a commutative diagram

$$
\begin{array}{ccccccccc}
0 & \longrightarrow & T_V \Phi F(n) & \longrightarrow & T_V F(n) & \longrightarrow & T_V \Sigma \Omega F(n) & \longrightarrow & 0 \\
& & \downarrow & & \downarrow & & \downarrow{\scriptstyle\cong} & & \\
0 & \longrightarrow & \Phi T_V F(n) & \longrightarrow & T_V F(n) & \longrightarrow & \Sigma \Omega T_V F(n) & \longrightarrow & 0
\end{array}
$$

The bottom row is obviously exact. Exactness of the top row follows from Proposition 2.2.3 and exactness of T_V. The right hand side vertical map is an isomorphism by 3.3.3 and 3.3.4. If one uses that T_V is exact, one knows that the top row is exact. It follows that $T_V \Phi F(n)$ is isomorphic to $\Phi T_V F(n)$, and we are done.

However, as one can expect, it is not necessary to use exactness of T_V. Recall that it is enough to check the proposition for the unstable $\mathcal{A}$-modules $F(n)$, i.e. to compute $T_V \Phi(F(n))$. It is also enough to consider the case $V = \mathbb{F}_p$, the general case follows by iteration.

Assume that $p = 2$. The group $\mathrm{Hom}_{\mathcal{U}}(\Phi F(n), N))$ identifies with the subgroup of $\mathrm{Hom}_{\mathcal{U}}(F(2n), N)) \cong N^{2n}$ of elements which are killed by any operation of odd degree. Indeed this follows from the observation that the kernel of the canonical surjective map from $F(2n)$ to $\Phi F(n)$ is generated, as an $\mathbb{F}_p$-vector space, by those classes $\mathrm{Sq}^I \iota_{2n}$ such that I involves at least one odd entry.

Then one computes $\mathrm{Hom}_{\mathcal{U}}(\Phi F(n), H^* \mathbb{Z}/2 \otimes N))$. Consider an $\mathcal{A}$-linear map $\varphi : \Phi F(n) \to H^* \mathbb{Z}/2 \otimes N$. The element $Q_i \Phi i_n$ is zero for any i. (Q_i is a Milnor derivation (3.6)). Also the element $\theta \Phi \iota_n$ is zero for any operation θ of odd degree. Consequently the same is true for $\theta \varphi \Phi \iota_n$, and in particular for $Q_i \varphi \Phi \iota_n$. Write $\varphi \Phi \iota_n$ as $\Sigma_\ell u^{2n-\ell} \otimes j_\ell$, $j_\ell \in N$. From the equation $Q_i \varphi \Phi \iota_n = 0$ for i large one gets:

(i) $j_\ell = 0$ if $\ell \equiv 1$ (2). From the equation $\theta \varphi \Phi i_n = 0$ for all θ of odd degree one gets:

(ii) $\theta j_{2\ell} = 0$ for all θ of odd degree (any ℓ).

The second point is proved by induction on $|\theta|$.

Therefore, the group $\mathrm{Hom}_{\mathcal{U}}(\Phi F(n), H^* \mathbb{Z}/2 \otimes N))$ identifies, naturally, with $\oplus_{i=0,\ldots,n} \mathrm{Hom}_{\mathcal{U}}(\Phi F(i), N))$ and the result follows using Lemma 3.3.1.

The case of an odd prime is left as an exercise to the reader. What precedes gives a direct computation of $T \Phi(F(n)) \to T(F(n))$ and that one does not need (3.4.2).

3.5. Commutation of T_V with tensor products

The main result of this section is

Theorem 3.5.1 [L1]. *For any unstable modules M_1 and M_2, there is a natural isomorphism*

$$\nu_{M_1,M_2} : T_V(M_1 \otimes M_2) \to T_V(M_1) \otimes T_V(M_2).$$

Let M_1, M_2, L_1, L_2 be in $\mathcal{U}$. We construct a natural map

$$\nu^{L_1,L_2}_{M_1,M_2} : (M_1 \otimes M_2 : L_1 \otimes L_2)_\mathcal{U} \to (M_1 : L_1)_\mathcal{U} \otimes (M_2 : L_2)_\mathcal{U}$$

as follows. Consider the maps $\varphi_i : M_i \to L_i \otimes (M_i : L_i)_\mathcal{U}$ $(i = 0, 1)$ adjoint to the identity of $(M_i : L_i)_\mathcal{U}$. Then the adjoint of $(1 \otimes \tau \otimes 1) \circ (\varphi_1 \otimes \varphi_2)$, where τ is the obvious twisting map, is by definition $\nu^{L_1,L_2}_{M_1,M_2}$.

If $L_1 = L_2 = L$ is an unstable algebra, the multiplication $\mu : L \otimes L \to L$ defines a map $\mu^*_M : (M : L)_\mathcal{U} \to (M : L \otimes L)_\mathcal{U}$. If $L = H^*V$, one defines ν_{M_1,M_2} to be the composite $\nu^{H^*V,H^*V}_{M_1,M_2} \circ \mu^*_{M_1 \otimes M_2}$.

Proof of Theorem 3.5.1. The proof that we shall give although close to the original proof of Lannes of [L1] does not use exactness of T. For yet another proof we refer to [L4].

Observe first that it is enough to work with $V = \mathbb{F}_p$.

As T is right exact (being a left adjoint), if $\nu_{F(p),F(q)}$ is an isomorphism, for all p and q, one concludes that $\nu_{M_1,F(q)}$ is an isomorphism for all $M \in \mathcal{U}$, all q. It follows, similarly, that ν_{M_1,M_2} is an isomorphism for all $M_1, M_2 \in \mathcal{U}$.

It remains to show that $\nu_{F(p),F(q)}$ is an isomorphism for all p, all q. Using the two exact sequences

$$0 \to F(p) \otimes \Phi F(q) \to F(p) \otimes F(q) \to F(p) \otimes \Sigma F(q-1) \to 0$$

$$0 \to \Phi F(p) \otimes \Phi F(q) \to F(p) \otimes \Phi F(q) \to \Sigma F(p-1) \otimes \Phi F(q) \to 0$$

one gets two commutative diagrams — in which we momentarily abbreviate $F(p)$ to F_p and $F(q)$ to F_q:

(3.5.2)

$$\begin{array}{ccccc}
T(F_p \otimes \Phi F_q) & \longrightarrow & T(F_p \otimes F_q) & \longrightarrow\!\!\!\!\rightarrow & T(F_p \otimes \Sigma F_{q-1}) \\
\downarrow & & \downarrow & & \downarrow \\
T(F_p) \otimes T(\Phi F_q) & \hookrightarrow & T(F_p) \otimes T(F_q) & \longrightarrow\!\!\!\!\rightarrow & T(F_p) \otimes T(\Sigma F_{q-1})
\end{array}$$

and

(3.5.3)

$$\begin{array}{ccccc}
T(\Phi F_p \otimes \Phi F_q) & \longrightarrow & T(F_p \otimes \Phi F_q) & \longrightarrow\!\!\!\!\rightarrow & T(\Sigma F_{p-1} \otimes \Phi F_q) \\
\downarrow & & \downarrow & & \downarrow \\
T(\Phi F_p) \otimes T(\Phi F_q) & \hookrightarrow & T(F_p) \otimes T(\Phi F_q) & \longrightarrow\!\!\!\!\rightarrow & T(\Sigma F_{p-1}) \otimes T(\Phi F_q)
\end{array}$$

In both diagrams, the bottom row is exact by inspection, using properties of T with respect to Σ and Φ. The top rows are exact as T is right exact. If one uses exactness of T, one gets also exactness on the left and we are done. However we now show the result follows without using exactness of T. The right hand side vertical maps can be assumed to be isomorphisms by an inductive hypothesis, and 3.3.4 (the case of the second one needs a small complementary argument which is left to the reader).

Now let us show that $\nu_{F(p),F(q)}$ is an isomorphism. The proof splits in two parts. Assume for the moment that p is 2.

3.5.4. $\nu_{F(p),F(q)}$ *is an isomorphism in degrees strictly greater than zero.*

This splits again in two parts: one shows that $\nu_{F(p),F(q)}$ is injective, then that it is surjective.

Let us prove injectivity. Let $x \neq 0$ be such that $\nu_{F(p),F(q)} x = 0$. Then there exists a non-zero element $x' \in T(F(p) \otimes \Phi F(q))$ such that $\nu_{F(p),\Phi F(q)} x' = 0$ (by 3.5.2). Then (3.5.3) implies that there exists a non-zero element $x'' \in T(\Phi F(p) \otimes \Phi F(p))$ such that $\nu_{\Phi F(p),\Phi F(q)} x'' = 0$. The element x'' of $T(\Phi F(p) \otimes \Phi F(q)) \cong \Phi T(F(p) \otimes F(q))$ can be written as Φx_1, $x_1 \in T(F(p) \otimes F(q))$ is such that $\mathrm{Sq}_0\, x_1 = x$. As $T(F(p)) \otimes T(F(q))$ is reduced,

$v_{F(p),F(q)}x_1$ has to be zero. As $2|x_1| = |x|$, an obvious induction shows that x has to be zero, which is a contradiction.

For the surjectivity we will give two proofs. First, we give a direct proof of surjectivity of the map:

$$v_{p,q} : T(F(p) \otimes F(q)) \to T(F(p)) \otimes T(F(q))$$

in strictly positive degrees. Again assume that we have isomorphisms

$$T(F(k) \otimes F(\ell)) \to T(F(k)) \otimes T(F(\ell)),$$
$$T(F(k) \otimes \Phi F(\ell)) \to T(F(k)) \otimes T(\Phi F(\ell)),$$
$$T(\Phi F(k) \otimes \Phi F(\ell)) \to T(\Phi F(k)) \otimes T(\Phi F(\ell)),$$

as soon as $k + \ell < p + q$.

Suppose that there is a non-zero element x_0 which is not in the image of $v_{p,q}$. Using (3.5.2) one observes that it is possible to assume that x_0 is the image of an element x_0' of $T(F(p)) \otimes T(\Phi F(q))$ which does not come from $T(F(p) \otimes \Phi F(q))$. Using (3.5.3) one observes that this implies, that there exists $x_1' \in T(\Phi F(p)) \otimes T(\Phi F(q))$ not in the image of $T(\Phi F(p) \otimes \Phi F(q)) \cong \Phi T(F(p) \otimes F(q))$. As the element x_1' can be written as Φx_1 for some x_1 which is not in the image of $v_{p,q}$, and as $|x_1| = |x_0|/2$, an induction leads to a contradiction. Note that this type of argument has been used, for example, in Chapter 2.

A 'dimension type' argument proving 3.5.4 will be given in 3.7.

3.5.5. $v_{F(p),F(q)}$ *is an isomorphism in degree zero.*

The unstable $\mathcal{A}$-module $T(F(p)) \otimes T(F(q))$ being of rank one in degree zero, it is enough to show that $T(F(p) \otimes F(q))$ is of rank one in degree zero, and that the map $v_{F(p),F(q)}$ is surjective in degree zero. The second property is easily checked once the first one is proved, it is left as an exercise to the reader (see Section 3.7). The first property depends on

Lemma 3.5.6. *For k large enough $T(F(p) \otimes \Phi^k F(q))$ is of dimension 1 in degree zero.*

This gives the result because the exact sequence

$$T(F(p) \otimes \Phi^\ell F(q)) \to T(F(p) \otimes \Phi^{\ell-1} F(q))$$
$$\to T(F(p) \otimes \Sigma\Omega\Phi^{\ell-1} F(q)) \to 0$$

reduces in degree zero to a surjection (in fact a bijection):

$$T(F(p) \otimes \Phi^\ell F(q))^0 \to T(F(p) \otimes \Phi^{\ell-1} F(q))^0 \to 0.$$

It remains to prove the lemma. We shall prove that, if k is large enough, the unstable $\mathcal{A}$-module $F(p) \otimes \Phi^k F(q)$ is monogenic. This implies the lemma because $T(F(p) \otimes \Phi^k F(q))$ is easily shown to be of dimension at least one.

Let $\theta\iota_p \otimes \theta' \operatorname{Sq}_0^k \iota_q$ be any class in $F(p) \otimes \Phi^k F(q)$. Observe that one can suppose that θ' is in the subalgebra of $\mathcal{A}$ generated by the operations Sq^{2^i} with $i \geq k$. Indeed (2.2.1) implies that any operation Sq^{2^i}, with $i < k$, acts trivially on any class of the form $\operatorname{Sq}_0^k x$. More generally let γ be an operation such that $|\gamma|$ is not divisible by 2^k, (2.2.1) shows that γ acts trivially on any class of the form $\operatorname{Sq}_0^k x$.

Assume that 2^k is strictly greater than p. An easy computation, using the preceding remark and the instability condition, shows that $\theta'(\iota_p \otimes \operatorname{Sq}_0^k \iota_q) = \iota_p \otimes \theta' \operatorname{Sq}_0^k \iota_q$.

Now if the operation θ is non-trivial on ι_p, one can suppose it is in the left ideal generated by the operations Sq^{2^i} for $2^i \leq p$. One observes that for such an operation, $\theta(\iota_p \otimes \operatorname{Sq}_0^k x) = \theta\iota_p \otimes \operatorname{Sq}_0^k x$ for any class $\operatorname{Sq}_0^k x$ (use again (1.7.1)), we assume still that $2^k > p$. Consequently we have $\theta\theta'(\iota_p \otimes \operatorname{Sq}_0^k \iota_q) = \theta\iota_p \otimes \theta' \operatorname{Sq}_0^k \iota_q$, and we are done.

3.6. The case of an odd prime

As usual the proof works 'exactly' the same way when one considers the category $\mathcal{U}'$ and replaces $H = H^*\mathbb{Z}/p$ by $P = H^*\mathbb{CP}^\infty$ and defines $T'M$ to be $(M : P)_{\mathcal{U}'}$.

Lemma 3.6.1 [L4]. *For any M in $\mathcal{U}'$, the natural map $TM \to T'M$ is an isomorphism.*

The 'natural map' is defined as follows. One considers the adjoint of the identity of $T'M$

$$M \to P \otimes T'M ,$$

composes it with $P \otimes T'M \hookrightarrow H^*\mathbb{Z}/p \otimes T'M$, and adjoints back.

To prove the lemma, it is enough to check it on the $F'(2n)$'s, and the only non-formal computation is that of $T(F'(2k))$.

It is enough to prove that for any map $\varphi : F'(2k) \to H^*\mathbb{Z}/p \otimes N$ (any N in $\mathcal{U}$) the element $\varphi(\iota'_{2k})$ can be written as $\sum_{0 \le i \le k} x^i \otimes n_i$, where the $\mathcal{A}$-module $\mathcal{A}n_i$ is an object in $\mathcal{U}'$ for all i. Indeed, this implies that $T(F'(2k))$ is isomorphic to $\oplus_{1 \le i \le k} F'(2i)$. The result follows. We note that there is a more general result. Let $\tilde{\mathcal{O}}$ be the right adjoint of the forgetful functor $\mathcal{O} : \mathcal{U}' \to \mathcal{U}$. Next let M and N be in $\mathcal{U}$, and assume that $\tilde{\Sigma}M = 0$. Then [LZ1]:

$$(3.6.2) \qquad \tilde{\mathcal{O}}(M \otimes N) \cong \tilde{\mathcal{O}}M \otimes \tilde{\mathcal{O}}N .$$

Our claim is the case where $M = H^*\mathbb{Z}/p$. We prove this case now. A priori $\varphi(\iota'_{2k})$ has the form $\sum_i x^i \otimes n'_i + \sum_j tx^{j-1} \otimes n_j$. As $\beta\varphi(\iota'_{2k})$ is trivial one gets $n_i = \beta n'_i$. Therefore, $\varphi(\iota'_{2k})$ has the form $\sum_i x^{i-1}(x \otimes n'_i + t \otimes \beta n'_i)$. Next one shows that $\beta n'_i = 0$ for any i. This uses derivations in the mod p Steenrod algebra. Namely consider the operations P_k^0 dual to the classes ξ_k in the monomial basis (see 1.10). As $\beta P_k^0 \varphi(\iota'_{2k})$ is trivial we get for any k the equation:

$$\sum_i (ix^{i+p^k-1} \otimes \beta n'_i + x^i \otimes \beta P_k^0 n'_i + x \otimes P_k^0 \beta n'_i - t \otimes \beta P_k^0 \beta n'_i)) = 0 .$$

By assuming that k is large enough this implies that $\beta n'_i$ is trivial as soon as i is prime to p. Replacing the operations P_k^0 by the operations which are dual to the classes ξ_k^p one gets similarly that $\beta n'_i$ is trivial as soon as p^2 does not divide i; and so on. Let P^I be an admissible monomial in $\mathcal{A}$ such that $\varepsilon_j = 0$ for all j (i.e. P^I does not involve β). We have to show that for any I and i one has $\beta P^I n'_i = 0$. One does it as above by looking at equations $\beta Q_k P^I \varphi(\iota'_{2k}) = 0$ for k large enough. Once again one has also to use operations dual to the elements $\xi_{k+1}^{p^h}$.

Here is alternate proof suggested by Kuhn. Applying the operation Q_k (see 1.10), k large, to $\sum_i x^{i-1}(x \otimes n'_i + t \otimes \beta n'_i)$, one gets directly that $\beta n_i' = 0$ for all i. Then one gets that $Q_k n_i' = 0$ for all i and k. Then using Theorem 4a of [**Mn**] one concludes easily that $\mathcal{A} n_i' \in \mathcal{U}'$.

Note also that if M is in $\mathcal{U}'$ $T(\Sigma M) \cong \Sigma T(M) \cong \Sigma T'(M)$.

We now prove Theorem 3.5.1 in the case of an odd prime. By a colimit argument, it is possible to assume M and N finitely generated over $\mathcal{A}$. Assume that M is in $\mathcal{U}'$ and N is in $\mathcal{U}$. The unstable module M has a finite filtration $N = N_t \supset N_{t-1} \supset \ldots \supset N_0 = 0$ such that, for any i, N_i / N_{i-1} is either in $\mathcal{U}'$ or is the suspension of an object in $\mathcal{U}'$. As the analogue of Theorem 3.5.1 holds for T', and because of Lemma 3.6.1 and the above remark one gets

$$T(M \otimes N_i / N_{i-1}) \cong T(M) \otimes T(N_i / N_{i-1}) .$$

A diagram chasing (that does not use left exactness of T) shows that the map $T(M \otimes N) \to T(M) \otimes T(N)$ is surjective. The general case follows. The proof given in 3.5.5 and 3.5.6 extends to $p > 2$. In this case, it is enough to observe that a map $\varphi : F(p) \otimes F(q) \to H^*\mathbb{Z}/p$ is determined by its restriction to $F(p)_p \otimes F(q)_q$ (see the proof 1.8.1 in the case of an odd prime for the definition); and then to observe that there is — up to scalar multiplication — just one non-trivial map from $F(p)_p \otimes F(q)_q$ to $H^*\mathbb{Z}/p$ (along an argument analogous to 3.5.6).

3.7. Comment and exercise

We describe below the promised dimension type argument to prove the surjectivity of $\nu_{p,q}$ in strictly positive degrees. This argument leads to a proof of Theorem 3.5.1, that does not depend on the exactness of T, or on the commutation of T with Φ (we leave this as an exercise for the reader).

Again let $p = 2$. For this argument, it is easier to dualize. One considers the following map (dual to $\nu_{F(p), F(q)}$ in degree d):

$$(3.7) \quad \bigoplus_{h+i=d} \mathrm{Hom}_{\mathcal{U}}(F(p), H \otimes J(h)) \otimes \mathrm{Hom}_{\mathcal{U}}(F(q), H \otimes J(i))$$

$$\longrightarrow \mathrm{Hom}_{\mathcal{U}}(F(p) \otimes F(q), H \otimes J(d)) .$$

In the preceding formula $H^*\mathbb{Z}/2$ is abbreviated H.

A basis for the left hand side is described as follows. Let $\mathcal{E}_{k,h,a}$, $1 \leq a \leq \dim F(k)^h$ be a basis of $F(k)^{h*} \cong \mathrm{Hom}_{\mathcal{U}}(F(k), J(h))$. Then the maps

$$\widetilde{\mathcal{E}}_{k,h;a} = u^{p-k} \otimes \mathcal{E}_{k,h,a} \circ (\iota_{p-k} \otimes \iota_k) : F(p) \to F(p-k) \otimes F(k) \to H \otimes J(h) ,$$

$0 \leq k \leq p$ and $1 \leq a \leq \dim F(k)^h$ form a basis of the vector space $\mathrm{Hom}_{\mathcal{U}}(F(p), H \otimes J(h))$. Let us note that in the definition of $\widetilde{\mathcal{E}}_{k,h,a}$ we have identified in the usual way u^{n-k} (resp. $\iota_{p-k} \otimes \iota_k$) with maps.

One gets the basis we are looking for by considering the tensor products

$$\widetilde{\mathcal{E}}_{k,h,a} \otimes \widetilde{\mathcal{E}}_{\ell,i,b} ,$$

with $0 \leq k \leq p$, $0 \leq \ell \leq q$, $1 \leq a \leq \dim F(k)^h$, $1 \leq b \leq \dim F(\ell)^i$ and $h, i \geq 0$ such that $h + i = d$.

The map (3.7) sends $\widetilde{\mathcal{E}}_{k,h,a} \otimes \widetilde{\mathcal{E}}_{\ell,i,b}$ to the following map

$$F(p) \otimes F(q) \xrightarrow{\widetilde{\mathcal{E}}_{k,h,a} \otimes \widetilde{\mathcal{E}}_{\ell,i,b}} H \otimes J(h) \otimes H \otimes J(i)$$

$$\xrightarrow{1 \otimes T \otimes 1} H \otimes H \otimes J(h) \otimes J(i) \xrightarrow{m \otimes \mu} H \otimes J(d)$$

where m is the multiplication, $\mu : J(h) \otimes J(i) \to J(h+i) = J(d)$ is the non-trivial map. We shall denote this map by $\widetilde{\mathcal{E}}_{k,h,a} * \widetilde{\mathcal{E}}_{\ell,i,b}$. It is enough to show that all these maps are linearly independent. This is proved as follows. Denote by $e_{k,h;a}$, $1 \leq a \leq \dim F(k)^h$ the basis of $F(k)^h$ dual to $\{\mathcal{E}_{k,h,a}\}_{1 \leq a \leq \dim F(k)^h}$. Recall that $d = h + i$ is fixed and let N be an integer large enough with respect to d. Consider the class $\mathrm{Sq}_0^N \iota_{p-k} \otimes e_{k,h,a}$ of $F(p-k) \otimes F(k) \subset F(1)^{\otimes p}$. Let $z_{k,h,a}^N = \Sigma_{\sigma \in \mathfrak{S}_p / \mathfrak{S}_{p-k} \times \mathfrak{S}_k} \sigma(\mathrm{Sq}_0^N \iota_{p-k} \otimes e_{k,h,a})$ be the corresponding symmetrized element in $F(p) \subset F(1)^{\otimes p}$. It is easy to check that if N is large enough with respect to d, then:

(i) $\widetilde{\mathcal{E}}_{k,h,a} * \widetilde{\mathcal{E}}_{\ell,i,b}(z_{k,h,a}^N \otimes z_{\ell,i,b}^N) = u^{N(p+q-\ell-k)} \otimes b_d$, where b_d is the generator of $J(d)^d$, and

(ii) $\widetilde{\mathcal{E}}_{k,h,a} * \widetilde{\mathcal{E}}_{\ell,i,b}(z_{k',h',a'}^N \otimes z_{\ell',i',b'}^N) = 0$ as soon as either $k' > k$ or $\ell' > \ell$ and as soon as $k = k'$, $\ell = \ell'$, and either $h' > h$ or $i' > i$.

These two facts easily imply the result.

3.8. The functor T_V and unstable algebras

In this section we show that if K is an unstable algebra $T_V K$ is in a natural way an unstable algebra and we derive some consequences:

Theorem 3.8.1 [L4]. *Let K, L be unstable $\mathcal{A}$-algebras. The unstable $\mathcal{A}$-module $T_V(K)$ is in a natural way an unstable $\mathcal{A}$-algebra and there is a natural isomorphism*

$$\mathrm{Hom}_{\mathcal{K}}(K, H^*V \otimes L) \cong \mathrm{Hom}_{\mathcal{K}}(T_V(K), L).$$

To prove this statement, we need to show:

(i) that the functor $\mathcal{K} \to \mathcal{K}$, $L \mapsto H^*V \otimes L$ has a left adjoint $K \mapsto (K : H^*V)_{\mathcal{K}}$ (again this is a 'division functor'); and

(ii) that, $(K : H^*V)_{\mathcal{K}}$ identifies with $T_V(K)$.

Here is the first proposition:

Proposition 3.8.2. *Let C be an unstable $\mathcal{A}$-algebra of finite type. The functor $L \mapsto L \otimes C$ from the category $\mathcal{K}$ to itself has a left adjoint $K \mapsto (K : C)_{\mathcal{K}}$. For any K, L in $\mathcal{K}$, there is a natural isomorphism*

$$\mathrm{Hom}_{\mathcal{K}}(K, C \otimes L) \cong \mathrm{Hom}_{\mathcal{K}}((K : C)_{\mathcal{K}}, L).$$

As in Chapter 6, this is implied by Freyd's adjoint functor theorem. However, we describe a proof similar to the one of 3.2.1. As we are working in a 'non-abelian', even non-additive setting, we need non-abelian resolutions. They are constructed using comonads (this was one reason to define comonads above).

This is the opportunity to define the comonad G on the category $\mathcal{K}$. The forgetful functor $\mathcal{O} : \mathcal{K} \to \mathcal{E}$ has a left adjoint $\mathcal{G} : \mathcal{E} \to \mathcal{K}$. The functor $\mathcal{G}$ is the composite, of the left adjoint $\mathcal{F}$ of the forgetful functor $\mathcal{U} \to \mathcal{E}$, and of the functor $U : \mathcal{U} \to \mathcal{K}$ of Steenrod-Epstein [SE]. The functor U is itself left adjoint to the forgetful functor $\mathcal{K} \to \mathcal{U}$. Therefore, for any $K \in \mathcal{K}$ and $M \in \mathcal{U}$, one has

(3.8.3) $\mathrm{Hom}_{\mathcal{U}}(M, \mathcal{O}K) \cong \mathrm{Hom}_{\mathcal{K}}(U(M), K).$

The unstable $\mathcal{A}$-algebra $U(M)$ is described as follows. Consider the symmetric algebra $S(M)$ which is the quotient of the tensor algebra $T(M)$ by the ideal generated by the elements $x \otimes y - (-1)^{|x||y|} y \otimes x$ for all x and y in M. This ideal is stable under the action of the Steenrod algebra. Therefore, $S(M)$ is an unstable $\mathcal{A}$-module and has a product for which the Cartan formula holds. However it is not true that, for y in M, one has $P_0 y = y^{\otimes p}$ in $S(M)$. Therefore, $U(M)$ is defined to be the quotient of $S(M)$ by the ideal generated by the elements $P_0 y - y^{\otimes p}$ for any $y \in M$. This ideal is stable under $\mathcal{A}$ and $U(M)$ is obviously an unstable $\mathcal{A}$-algebra. It is nothing other than the universal enveloping algebra of M regarded as an abelian restricted Lie algebra. The functor $M \mapsto U(M)$ is clearly left adjoint to the forgetful functor $\mathcal{K} \to \mathcal{U}$.

Let $\mathcal{G}$ be the functor $U \circ \mathcal{F} : \mathcal{E}_{\mathrm{gr}} \to \mathcal{K}$ and G be $\mathcal{G} \circ \mathcal{O}$; we have a comonad (G, η, ν) on the category $\mathcal{K}$. The unstable $\mathcal{A}$-algebras, of the form $\mathcal{G}(E)$ for some E in $\mathcal{E}_{\mathrm{gr}}$, are the 'free' unstable $\mathcal{A}$-algebras.

Note that if E is a connected graded $\mathbb{F}_2$-vector space ($E^0 = 0$), then $\mathcal{G}(E)$ is a symmetric algebra (polynomial if $p = 2$) on the graded vector space of generators $\Sigma \Omega \mathcal{F}(E)$ (see [BK1] for example). This will be important in the sequel (Chapter 7).

To any unstable $\mathcal{A}$-algebra K we associate a simplicial unstable $\mathcal{A}$-algebra $G_\bullet(K)$ by the formulas:

— $G_\bullet(K)_n = G^{n+1}(K)$;

— $d_i : G_\bullet(K)_n \to G_\bullet(K)_{n-1}$ is given by $d_i = G^i(\eta_{G^{n-i}(K)})$;

— $s_j : G_\bullet(K)_n \to G_\bullet(K)_{n+1}$ is given by $s_j = G^j(\nu_{G^{n-j}(K)})$.

Recall that in Section 3.2 we defined the division functor in the category $\mathcal{E}_{\mathrm{gr}}$. Define now $(\mathcal{G}(E) : C)_{\mathcal{K}}$ for E in $\mathcal{E}_{\mathrm{gr}}$ as

$$G((E : \mathcal{O}C)_{\mathcal{E}_{\mathrm{gr}}}).$$

This definition makes sense in view of the adjunction formulas. Then, as in Section 3.2, it is necessary to define the map induced by any unstable $\mathcal{A}$-algebra map $f : \mathcal{G}(E) \to \mathcal{G}(E')$. This is done as in Section 3.2 and left to the reader. Then the following is an exact sequence of unstable $\mathcal{A}$-modules for any $K \in \mathcal{K}$:

$$\to G_\bullet(K)_1 \xrightarrow{d_0 - d_1} G_\bullet(K)_0 \xrightarrow{\varepsilon_K} K \to 0.$$

The unstable $\mathcal{A}$-algebras $G_\bullet(C)_1$ and $G_\bullet(C)_0$ are free and we can define $(K : C)_{\mathcal{K}}$ to be the cokernel of

$$(G_\bullet(C)_1 : C)_{\mathcal{K}} \xrightarrow{(d_0:C)_{\mathcal{K}} - (d_1:C)_{\mathcal{K}}} (G_\bullet(K)_0 : C)_{\mathcal{K}} .$$

As defined, $(K : C)_{\mathcal{K}}$ is an unstable $\mathcal{A}$-algebra, because the maps $(d_0 : C)_{\mathcal{K}}$ and $(d_1 : C)_{\mathcal{K}}$ are unstable $\mathcal{A}$-algebra maps, and because of the simplicial identities $d_0 s_0 = d_1 s_0 = \mathrm{Id}$. The rest of the proof goes as in Section 3.2.

We now prove

Proposition 3.8.4 [L1]. *For any unstable $\mathcal{A}$-algebra K, the unstable $\mathcal{A}$-module $T_V(K)$ is in a natural way an unstable $\mathcal{A}$-algebra which is naturally isomorphic to $(K : H^*V)_{\mathcal{K}}$.*

We start by showing that $T_V(K)$ is in natural way an unstable $\mathcal{A}$-algebra. One defines a product $T_V(K) \otimes T_V(K) \to T_V(K)$ as the composite

$$T_V(K) \otimes T_V(K) \cong T_V(K \otimes K) \xrightarrow{T_V(m)} T_V(K) .$$

One checks easily that this defines on $T_V(K)$ a structure of $\mathbb{F}_p$-algebra with unit which is both associative and commutative.

That the Cartan formula holds is easy to check; only axiom $(\mathcal{K}2)$ needs some care. Consider the case $p = 2$. The axiom $(\mathcal{K}2)$ says that the two maps:

$$\Phi K \xrightarrow{\lambda_K} K \quad \text{and} \quad \Phi K \xrightarrow{\iota_K} (K \otimes K)_{\mathfrak{S}_2} \xrightarrow{\tilde{m}_K} K ,$$

where $\iota_K(\Phi k) = k \otimes k$ and $\tilde{m}_K$ is induced by the product, are the same. The map $T_V(\lambda_K)$ identifies with $\lambda_{T_V K}$ modulo the equivalence $\Phi T_V \cong T_V \Phi$ (see Section 3.4). By the very definition $T_V(\tilde{m}_K) = \tilde{m}_{T_V K}$. Finally, we leave it to the reader to check that ι_K is $\mathcal{A}$-linear and that $T_V(\iota_K) = \iota_{T_V K}$ (it is easier to work with the dual maps). The result follows.

We now check that, with this structure, $T_V K$ is isomorphic to $(K : H^*V)_{\mathcal{K}}$. Let L be in $\mathcal{K}$. Consider the 'diagram'

$$\mathrm{Hom}_{\mathcal{K}}(K, H^*V \otimes L) \longleftarrow \mathrm{Hom}_{\mathcal{U}}(K, H^*V \otimes L)$$

$$\Big\downarrow{\cong}$$

$$\mathrm{Hom}_{\mathcal{K}}(T_V K, L) \longleftarrow \mathrm{Hom}_{\mathcal{U}}(T_V K, L)$$

The definition of the product on $T_V K$ implies that the adjunction isomorphism sends $\mathrm{Hom}_{\mathcal{K}}(K, H^*V \otimes L)$ into $\mathrm{Hom}_{\mathcal{K}}(T_V K, L)$ (consider first the case where K is free). In the other direction, in order to show that the inverse image, under the adjunction isomorphism, of the subset $\mathrm{Hom}_{\mathcal{K}}(T_V K, L)$ of $\mathrm{Hom}_{\mathcal{U}}(T_V K, L)$ is contained in $\mathrm{Hom}_{\mathcal{K}}(K, H^*V \otimes L)$, it is enough to show that the adjoint map $K \to H^*V \otimes T_V K$ of the identity of $T_V K$ is a map of unstable $\mathcal{A}$-algebras. This is done by going back to the definition of T_V and to the definition of the product in $T_V(K)$.

Here is one computation of $T_V(K)$:

Lemma 3.8.5. *For any unstable $\mathcal{A}$-module M, the natural map $T_V \circ U(M) \to U \circ T_V(M)$ is an isomorphism of unstable $\mathcal{A}$-algebras.*

We describe the natural map $T_V \circ U(M) \to U \circ T_V(M)$. Consider the composite of the adjoint of $\mathrm{Id}_{T_V M}$, $M \to H^*V \otimes T_V M$, with the inclusion $H^*V \otimes T_V M \hookrightarrow H^*V \otimes U \circ T_V(M)$. It extends to an algebra map $U(M) \to H^*V \otimes U \circ T_V(M)$, and one takes the adjoint. That this is an isomorphism follows from the various adjonction isomorphisms.

Let us derive some important special cases of the preceding results.

A p-Boolean algebra is a commutative, unital, $\mathbb{F}_p$-algebra in which $x^p = x$ for any element x. The $\mathbb{F}_p$-vector space of degree zero elements of an unstable algebra form a p-Boolean algebra by Axiom $(\mathcal{K}_2)$. Therefore, for any unstable $\mathcal{A}$-algebra K, $T_V^0(K)$ is a p-Boolean algebra.

The spectrum of a p-Boolean algebra B is going to be denoted by $\mathrm{Spec}(B)$. Recall that the spectrum of a ring is the set of prime ideals which, in this case, identifies with the set of maximal ideals, and with the set of algebra maps from B to $\mathbb{F}_p$. As any p-Boolean algebra is colimit of its finite dimensional p-Boolean algebras $\mathrm{Spec}(B)$ is, in a natural way, a profinite set. The structure theorem for p-Boolean algebra says, that the canonical map from B to the p-Boolean algebra, of continuous map from the profinite set $\mathrm{Spec}(B)$ to $\mathbb{F}_p$, provided with the discrete topology is an isomorphism. Recall that the canonical map sends an element $x \in B$ on the map from $\mathrm{Spec}(B)$ to $\mathbb{F}_p$ which associates, to a maximal ideal $\mathcal{M}$, the class of x in the field $B/\mathcal{M} = \mathbb{F}_p$.

The theorem is proved in the usual way [K2]. One first considers the finite dimensional case. Let x be a non-zero element which is not equal to the unit. The algebra B is the (non-trivial) product of the ideals generated by x^{p-1} and $1 - x^{p-1}$. Each of this ideal is again a p-Boolean algebra with unit x^{p-1} and $1 - x^{p-1}$. Thus we are easily done by induction. From this point the proof is completed by a 'limit and colimit argument'.

An unstable $\mathcal{A}$-module M is the colimit of its finitely generated sub $\mathcal{A}$-modules $\{M_\lambda\}$, $\lambda \in \Lambda$. Therefore, $\mathrm{Hom}_{\mathcal{U}}(M, H^*V)$ is, naturally, the limit over Λ of the finite dimensional $\mathbb{F}_p$-vector spaces $\mathrm{Hom}_{\mathcal{U}}(M_\lambda, H^*V)$. This is a profinite $\mathbb{F}_p$-vector space. In particular, $T_V^0(M)$ identifies with the continuous dual $\mathrm{Hom}_{\mathcal{U}}(M, H^*V)'$. On the other hand, an unstable algebra K is the colimit of its finitely generated $\mathcal{A}$-subalgebras $\{K_\lambda\}$, $\lambda \in \Lambda$. Therefore, $\mathrm{Hom}_{\mathcal{K}}(K_\lambda, H^*V)$ is the limit over Λ of the finite sets $\mathrm{Hom}_{\mathcal{K}}(K_\lambda, H^*V)$ and is in a natural way a profinite set. The inclusion $\mathrm{Hom}_{\mathcal{K}}(K, H^*V) \to \mathrm{Hom}_{\mathcal{U}}(K, H^*V)$ is continuous by the definition of the profinite structure. Thus we get a restriction map

$$\mathrm{Hom}_{\mathcal{U}}(K, H^*V)' \cong T_V^0(K) \to \mathbb{F}_p^{\mathrm{Hom}_{\mathcal{K}}(K,H^*V)}$$

from $T_V^0(K)$ to the p-Boolean algebra of continuous maps from the profinite set $\mathrm{Hom}_{\mathcal{K}}(K, H^*V)$ to $\mathbb{F}_p$ (equipped with the discrete topology).

The following theorem of Lannes, [LZ2][L1], should be thought of as a 'linearization principle'

Theorem 3.8.6. *The preceding restriction map is an isomorphism of p-Boolean algebras.*

This is a map of p-Boolean algebras by the construction; that it is an isomorphism is proved using the structure theorem for p-Boolean algebras and Theorem 3.8.1 for $L = \mathbb{F}_p$. In fact Theorem 3.8.1 identifies $\mathrm{Spec}(T_V^0 K)$ with $\mathrm{Hom}_{\mathcal{K}}(K, H^*V)$.

Corollary 3.8.7. *(Non linear injectivity of H^*V) Let $i : K \to K'$ be an injection of unstable $\mathcal{A}$-algebras, $\varphi : K \to H^*V$ a map of unstable $\mathcal{A}$-algebras. Then there exists a map $\varphi' : K' \to H^*V$ of unstable $\mathcal{A}$-algebras such that $\varphi' \circ f = \varphi$.*

Proof. As T_V is exact, one has an injection of p-Boolean algebras

$$0 \to T_V^0(K) \to T_V^0(K'),$$

and a surjection

$$0 \leftarrow \operatorname{Hom}_{\mathcal{K}}(K, H^*V) \leftarrow \operatorname{Hom}_{\mathcal{K}}(K', H^*V).$$

This needs a proof. Indeed we are dealing with continuous maps. Suppose that we have an injection $B \to B'$ of p-Boolean algebras, we show that the induced map $\operatorname{Spec}(B') \to \operatorname{Spec}(B)$ is surjective. The image of $\operatorname{Spec}(B')$ is compact and can be separeated from any point that is not inthere. Thus if there is any point outside of the image there is a map from $\operatorname{Spec}(B)$ to $\mathbb{F}_p$ which separates this point and the image. It follows easily that $B \to B'$ is not injective, which is a contradiction.

Let $\mathcal{C}$ be a finite category (which means that at $\mathcal{C}$ has both finitely many objects and finitely many morphisms), and let $\gamma : \mathcal{C} \to \mathcal{K}$ be a contravariant functor.

Corollary 3.8.8 [L1]. *The natural map*

$$\operatorname{colim}_{\mathcal{C}} \operatorname{Spec} \circ T_V^0 \circ \gamma \to \operatorname{Spec} \circ T_V^0(\lim_{\mathcal{C}} \gamma)$$

is an isomorphism. In other words, the functor $K \mapsto \operatorname{Hom}_{\mathcal{K}}(K, H^*V)$, $\mathcal{K} \to$ (profinite sets) *transforms a finite limit into a colimit.*

Proof. The limit of the functor γ over $\mathcal{C}$ is defined to be the kernel of the map

$$\prod_{c \in \mathcal{C}} \gamma(c) \quad \overset{\ell}{\longrightarrow} \quad \prod_{\varphi \in \operatorname{Mor}(\mathcal{C})} \gamma(s\varphi),$$

where $s\varphi$ is the source of φ and the map ℓ sends the element (x_c), $c \in \mathcal{C}$ to the element $(x_{s\varphi} - \gamma(\varphi)x_{t\varphi})$, $\varphi \in \operatorname{Mor}(\mathcal{C})$ ($t\varphi$ is the target of φ). If $\mathcal{C}$ is the category defined by a finite group G (one object and one morphism for each element in G), a functor γ is just an unstable algebra equipped with a G-action and $\lim_{\mathcal{C}} \gamma$ identifies with the invariants K^G.

Now T_V, and hence T_V^0, commutes with finite products (which coincide with finite sums) and we get by exactness of T_V that $T_V^0(\lim_{\mathcal{C}} \gamma)$ is the kernel of

$$\prod_{c \in \mathcal{C}} T_V^0 \circ \gamma(c) \xrightarrow{\ T_V^0 \ell\ } \prod_{\varphi \in \operatorname{Mor}(\mathcal{C})} T_V^0 \circ \gamma(s\varphi) .$$

Therefore

$$T_V^0(\lim_{\mathcal{C}} \gamma) \to \lim_{\mathcal{C}}(T_V^0 \circ \gamma)$$

is an isomorphism of p-Boolean algebras. Considering the spectra, we get that

$$\operatorname{Spec} \circ T_V^0(\lim_{\mathcal{C}} \gamma) \leftarrow \operatorname{colim}_{\mathcal{C}} \operatorname{Spec} \circ T_V^0 \circ \gamma$$

is an isomorphism.

3.9. Further examples of T_V -computation

We recollect here a few examples of T_V -computation that appear in what precedes and give other ones concerning $T_V^0 K$.

The most basic example is the case of the mod p-cohomology of an elementary abelian p-group W:

(3.9.1) $T_V H^* W \cong H^* W^{\operatorname{Hom}(V,W)} .$

The degree zero part of this result is known as the Adams-Gunawardena-Miller theorem. After dualization it writes as

$$\operatorname{Hom}_{\mathcal{U}}(H^* W, H^* V) \cong \mathbb{F}_p[\operatorname{Hom}(V, W)] .$$

In fact the result of Adams, Gunawardena and Miller in [**AGM**] contains this one as a particular case. It should also be noted that this is also a particular case of Theorem 3.8.6. Recall that (Section 1.5) the bijection $\operatorname{Hom}_{\mathcal{K}}(H^* V, H^* W) \cong \operatorname{Hom}(W, V)$ is easy.

More generally if K of the form $U(M)$ for some unstable $\mathcal{A}$-module M, then

(3.9.2) $T_V U(M) \cong U(T_V M) .$

We also know that T_V commutes with Φ:

(3.9.3) $T_V \Phi K \cong \Phi T_V K .$

The following example uses the notion of locally finite $\mathcal{A}$-module. Recall that an $\mathcal{A}$-module M is said to be locally finite if for any

x in M the cyclic $\mathcal{A}$-module $\mathcal{A}x$ is finite (i.e. only finitely many operations act non-trivially on x). If K is an unstable $\mathcal{A}$-algebra which is locally finite as an $\mathcal{A}$-module then

$$(3.9.4) \qquad T_V K \cong K ,$$

for this is true for any locally finite unstable $\mathcal{A}$-module.

Let K be an unstable $\mathcal{A}$-algebra provided with the action of a finite group G. Then, by naturality, G acts on $T_V K$ and

$$(3.9.5) \qquad T_V(K^G) \cong (T_V K)^G .$$

This follows from the following more general result. Let $\mathcal{C}$ be a finite category and γ be a functor from $\mathcal{C}$ to $\mathcal{K}$. Then

$$(3.9.6) \qquad T_V(\lim_{\mathcal{C}} \gamma) \cong \lim_{\mathcal{C}} T_V \circ \gamma .$$

Next let K be an augmented unstable $\mathcal{A}$-algebra, $\varepsilon : K \to \mathbb{F}_p$ being the augmentation. The vector space $T_V^0 K$ is a p-Boolean algebra; let φ be an element in $\mathrm{Hom}_{\mathcal{K}}(K, H^*V)$. Denote by ε_φ the primitive idempotent of $T_V^0 K$ which is the function taking the value 1 on φ, and the value 0 anywhere else. Define $(T_V K)_\varphi$ to be $\varepsilon_\varphi T_V K$. It is easy to observe that

$$(T_V K)_\varphi \cong T_V K \otimes_{T_V^0 K} \mathbb{F}_p ,$$

where $\mathbb{F}_p$ is a $T_V^0 K$-module via the adjoint of φ which is the evaluation map at φ.

Of particular importance is the case of the adjoint of the trivial map $\tau : K \xrightarrow{\varepsilon} \mathbb{F}_p \xrightarrow{\eta} H^*V$, where η is the unit. We shall denote this adjoint by t. Recall the indecomposable functor $Q : \mathcal{K} \to \mathcal{U}$

$$Q(K) = I/I^2 ,$$

I being the kernel of the augmentation.

Proposition 3.9.7. *Let K be a connected ($K_0 \cong \mathbb{F}_p$) unstable $\mathcal{A}$-algebra such that $Q(K)$ is locally finite as an $\mathcal{A}$-module. Then $(T_V K)_\tau$ is isomorphic to K.*

Proof. There is a natural map $K \to (T_V K)_\tau$ obtained as follows. One considers the map $K \to T_V K$ induced by the map $H^*V \to \mathbb{F}_p$, it is the composite of the map $K \to H^*V \otimes T_V K$, which is the adjoint of the identity of $T_V K$, with the projection on $T_V K$. Then one projects onto $(T_V K)_\tau$.

On the other hand, the map $T_V K \to K$, which is adjoint to the inclusion $K \hookrightarrow H^*V \otimes K$, factors through $(T_V K)_\tau$. Therefore, we have maps $K \to (T_V K)_\tau$, and $(T_V K)_\tau \to K$ whose composite is easily shown to be the identity of K.

As both K and $(T_V K)_\tau$ are connected, it is enough to show that $Q(K) \cong Q[(T_V K)_\tau]$. But exactness of T_V and (3.5.1) imply that $Q[T_V(K)] \cong T_V[Q(K)]$, where $Q[T_V K]$ is computed using the augmentation $\tau = T_V(\varepsilon)$ of $T_V(K)$. But $Q[T_V(K)]$ is isomorphic to $Q[T_V(K)_\tau]$, by the definition, and $Q(K)$ is isomorphic to $Q[T_V(K)_\tau]$. The result follows (see also [Sc1]). We refer to [DW2] for another proof.

Consider now a subgroup G of the automorphism group $\mathrm{Aut}\, V$ of an elementary abelian p-group V. For convenience, we shall assume here that $\gamma \in G$ acts on H^*V by $x \to (\gamma^{-1})^*x$. Let K be the subalgebra of invariants $(H^*V)^G$, i the inclusion of K in H^*V and φ an element of $\mathrm{Hom}(W, V)$.

Proposition 3.9.8. *The unstable $\mathcal{A}$-algebra $T_W K$ is isomorphic to the (graded) set of G-equivariant maps from $\mathrm{Hom}(W, V)$ to H^*V. The component $(T_W K)_{\varphi^* \circ i}$ is isomorphic to the subalgebra of invariants $H^*V^{G_\varphi}$, where G_φ is the subgroup of G of elements γ such that $\gamma \circ \varphi = \varphi$.*

Note that if $W = V$ and φ is the identity, one finds $(T_V K)_{\mathrm{Id}} \cong H^*V$.

The proof is left to the reader as well as the extension to the case where $(H^*V)^G$ is replaced by the limit of a functor $H^* \circ \Gamma$, where Γ is a functor from a finite category to $\mathcal{E}$. Let us observe that if φ is the trivial map t we have $(T_W K)_{t^* \circ i} \cong K$ (which fits with Proposition 3.9.7). We refer to [DW1] for other computations.

3.10. The formula for $T_V H^* BG$

We describe in this section the computation of $T_V^0 H^* BG$ for a finite group G. We first recall a theorem of Quillen. Let $\mathcal{C}(G)$ be the category whose objects are the abelian p-subgroups E; the morphisms being subconjugacy relations (i.e. a morphism $E \to F$ is a morphism of groups induced by an interior automorphism of G). There is a natural map

$$q_G : H^* BG \to \lim_{\mathcal{C}(G)} H^* BE .$$

Quillen's theorem **[Q2]** says that q_G is an F-isomorphism, which means, that any element in the kernel of q_G is nilpotent, and that for any element y in the target, there exists n such that y^{p^n} is in the image. In our terminology this means that Ker q_G and Coker q_G are nilpotent unstable $\mathcal{A}$-modules. It is time to note.

Lemma 3.10.1. *Let M be a nilpotent unstable $\mathcal{A}$-module. Then $T_V^0 M$ is trivial.*

This is clear for $p = 2$, for the operation Sq_0 is injective on $H^* \mathbb{Z}/2$ and nilpotent on any element in M. For $p > 2$, it follows from the fact that $H^* V$ is reduced (2.6.5).

Let G be a finite group, denote by $\text{Rep}(V, G)$ the set $\text{Hom}(V, G)$ modulo conjugacy at the target. There is a canonical bijection

$$\text{colim}_{\mathcal{C}(G)} \ \text{Hom}(V, E) \to \text{Rep}(V, G) .$$

The following consequence of Quillen's theorem has been proved by various people, among them Adams, Lannes, Miller and Wilkerson.

Theorem 3.10.2. *The p-Boolean algebra $T_V^0 H^* BG$ is isomorphic to the p-Boolean algebra of functions from $\text{Rep}(V, G)$ to $\mathbb{F}_p$.*

One applies Lemma 3.10.1, exactness of T and Corollary 3.8.8, Example 3.9.1 and the remark preceding the theorem.

Let φ be an element of $\text{Rep}(V, G)$ and Z_φ the centralizer of φ in G

$$Z_\varphi = \{ g \in G / g\varphi(v)g^{-1} = \varphi(v) \ \ \forall v \in V \} .$$

The obvious homomorphism $V \times Z_\varphi \to G$ determines a map a_φ : $T_V H^* BG \to H^* BZ_\varphi$.

Lannes [L2] has computed $T_V H^* BG$; the result is as follows

Theorem 3.10.3. *The sum map*

$$\oplus_\varphi a_\varphi : T_V H^* BG \to \oplus H^* BZ_\varphi$$

is an isomorphism.

3.11. The classification theorem for injective unstable $\mathcal{A}$-modules

The object of the following sections is to show that we have described all the injective unstable $\mathcal{A}$-modules. More precisely, let us introduce a set of representatives of indecomposable summands of the unstable $\mathcal{A}$-modules $H^{\otimes m}$, $m \geq 0$ (by convention $H^{\otimes 0} \cong \mathbb{F}_p$), H denoting as usual $H^* \mathbb{Z}/p$. Denote this set by $\mathcal{L}$. Recall that, by set of representatives, one means that $\mathcal{L}$ contains one object in each isomorphism class of indecomposable factors of $H^{\otimes m}$, $m \geq 0$, and only one. Recall also that an indecomposable module is a module which is not a non-trivial direct sum.

Theorem 3.11.1 [LS2]. *Let I be an injective unstable $\mathcal{A}$-module. There exists a unique family of cardinals $(a_{L,n})$, $(L, n) \in \mathcal{L} \times \mathbb{N}$, such that*

$$I \cong \bigoplus_{(L,n)} (L \otimes J(n))^{\oplus a_{L,n}} .$$

Recall that $M^{\oplus a}$ denotes the direct sum of a-copies of M.

Recall also that as the category $\mathcal{U}$ is locally noetherian, a direct sum of injective objects is still injective (see 2.1.2).

The following proposition is implicit in the theorem:

Proposition 3.11.2. *The unstable $\mathcal{A}$-module $L \otimes J(n)$ is indecomposable for any $L \in \mathcal{L}$, any $n \in \mathbb{N}$.*

Observe that Theorem 3.11.1 implies that if $L \otimes J(n) \cong L' \otimes J(n')$ ($L, L' \in \mathcal{L}$ and $n, n' \in \mathbb{N}$), then $L = L'$ and $n = n'$.

Before proving (3.11.1), we need to recall generalities about injective objects in abelian categories (see [Br] or [Gb]) for details. Recall the notion of injective hull.

Definition 3.11.3. Let M be an unstable $\mathcal{A}$-module. An *injective hull* for M is a couple (I, i) where I is an injective unstable $\mathcal{A}$-module and $i : M \to I$ is an essential $\mathcal{A}$-linear map.

A map $\varphi : M \to N$ is said to be *essential* if a sub $\mathcal{A}$-module P of N is trivial if and only if $\varphi^{-1}(P)$ is trivial. In particular any essential map is injective.

Obviously, an injective hull, if it exists, is unique up to isomorphism.

Proposition 3.11.4. *In the category* $\mathcal{U}$, *injective hulls exist.*

This is true for any abelian category having a set of generators and exact filtered colimits ([Gb], chap.2, Th.2).

Recall some characterizations of indecomposable injective modules.

Proposition 3.11.5. *Let* I *be an injective unstable* $\mathcal{A}$-module. The following conditions are equivalent:

(i) I *is indecomposable;*

(ii) *if* M *and* N *are sub* $\mathcal{A}$-modules of I, and $M \cap N$ is trivial, then either M or N is trivial;*

(iii) I *is the injective hull of each of its non-trivial sub* $\mathcal{A}$-modules;*

(iv) $\mathrm{End}_{\mathcal{U}}(I)$ *is a local ring (i.e. f or $\mathrm{Id} - f$ is an isomorphism for any f in $\mathrm{End}_{\mathcal{U}}(I)$).*

The proofs are just the same as in the general case. We refer to [Br] or to Section 5 in [Gb].

As an example, observe that $J(n)$ is indecomposable since $\mathrm{End}_{\mathcal{U}}(J(n)) \cong \mathbb{F}_p$ is local. Thus this is the injective hull of $\Sigma^n \mathbb{F}_p$ [Ml2].

Corollary 3.11.6. *Let I be an injective unstable A-module which is indecomposable and i be an injection $I \hookrightarrow \oplus_{\lambda \in \Lambda} M_\lambda$. Then there exists $\lambda_0 \in \Lambda$ such that $\mathrm{pr}_{\lambda_0} \circ i$ is an injection (pr_λ denoting the projection).*

Proof. This follows immediately from (iii) of (3.13.5) if Λ is finite. If not, consider for a non-zero x in I, the finite set Λ_0 of those λ for which $\mathrm{pr}_\lambda \circ i(x)$ is non-zero. Let M' be the direct sum $\oplus_{\lambda \in \Lambda_0} M_\lambda$, M'' be the direct sum $\oplus_{\lambda \in \Lambda - \Lambda_0} M_\lambda$, pr' and pr'' the corresponding projections. As the intersection $\mathrm{Ker}\,\mathrm{pr}' \circ i \cap \mathrm{Ker}\,\mathrm{pr}'' \circ i$ is trivial, one gets by (3.13.5) that $\mathrm{Ker}\,\mathrm{pr}' \circ i = 0$ and one is done by the preceding case.

Denote by $\mathcal{J}$ a set of representatives of indecomposable injective unstable A-modules (see the beginning of the chapter for the meaning of 'set of representatives'). Such a set exists because any indecomposable $\mathcal{U}$-injective is the injective hull of a cyclic A-module, and because there exists obviously a set of representatives for such modules.

Then the Azumaya-Krull-Schmidt Theorem says that an injective object of a locally noetherian category is in a unique way a direct sum of injective indecomposable objects (see **[Br]** Theorem 3, p. 22 and **[Po]** Theorem 5.8.11). In our case:

Theorem 3.11.7. *Let I be an injective unstable A-module. There exists a unique family of cardinals (a_E), $E \in \mathcal{J}$, such that*

$$I \cong \bigoplus_E E^{\oplus a_E}.$$

We are left with the task of describing indecomposable injective unstable A-modules. We shall proceed in two steps, we start by determining the reduced indecomposable unstable A-modules and then we consider the general case.

3.12. Reduced indecomposable $\mathcal{U}$-injectives

Let I be a reduced indecomposable $\mathcal{U}$-injective. We want to show:

Theorem 3.12.1. *There exists an integer m for which I is a direct factor of $H^*(\mathbb{Z}/p^{\oplus m})$.*

Let us denote $\tilde{H}^*\mathbb{Z}/p$ by $\bar{H}$. By convention $\bar{H}^{\otimes 0}$ will be identified with $\mathbb{F}_p$. Theorem 3.12.1 is a consequence of:

Proposition 3.12.2. *For any positive integer i, there exists a family of integers $(a_{i,m})$, $m \in \mathbb{N}$, and an embedding of $\mathcal{A}$-modules*

$$K(i) \hookrightarrow \bigoplus_m (\bar{H}^{\otimes m})^{\oplus a_{i,m}} .$$

Let us prove Theorem 3.12.1. We know from Proposition 2.6.3 that I embeds in a product $\Pi_\alpha K(i_\alpha)$. If we assume that I is not isomorphic to $\mathbb{F}_p$, we may assume that $i_\alpha > 0$ for every α. Therefore, by Proposition 3.12.2, I embeds in the product

$$\prod_\alpha \bigoplus_m (\bar{H}^{\otimes m})^{\oplus a_{i_\alpha,m}} .$$

Hence, it is enough to check that $\prod_\alpha \bigoplus_m (\bar{H}^{\otimes m})^{\oplus a_{i_\alpha,m}}$ can be written as $\bigoplus_m (\bar{H}^{\otimes m})^{\oplus b_m}$ for certain cardinals b_m. Then Corollary 3.11.4 implies that I injects in $\bar{H}^{\otimes m}$ for some m. But $\prod_\alpha \bigoplus_m (\bar{H}^{\otimes m})^{\oplus a_{i_\alpha,m}}$ isomorphic to the colimit

$$\operatorname{colim}_k \prod_\alpha \bigoplus_{0 \le m \le k} (\bar{H}^{\otimes m})^{\oplus a_{i_\alpha,m}} .$$

This is true because, in a given degree, $\bar{H}^{\otimes m}$ is non-zero for only finitely many values of m. Denote by L_k the unstable $\mathcal{A}$-module $\prod_\alpha \bigoplus_{0 \le m \le k} (\bar{H}^{\otimes m})^{\oplus a_{i_\alpha,m}}$. We have the following isomorphism

$$L_k = \prod_\alpha \bigoplus_{0 \le m \le k} (\bar{H}^{\otimes m})^{\oplus a_{i_\alpha,m}} \cong \bigoplus_{0 \le m \le k} \prod_\alpha (\bar{H}^{\otimes m})^{\oplus a_{i_\alpha,m}} .$$

Observe that $\prod_\alpha (\bar{H}^{\otimes m})^{\oplus c_\alpha}$ is isomorphic to $\prod_\alpha \bar{H}^{\otimes m} \otimes \mathbb{F}_p^{\oplus c_\alpha}$ and to $\bar{H}^{\otimes m} \otimes \prod_\alpha \mathbb{F}_p^{\oplus c_\alpha}$ since products commute with finite direct sums. Therefore, to $(\bar{H}^{\otimes m})^{\oplus b}$ for a certain cardinal b. Therefore, the module L_k is isomorphic to a direct sum $\bigoplus_{0 \leq m \leq k}(\bar{H}^{\otimes m})^{\oplus b_m}$ for certain cardinals b_m. Thus we get

$$\prod_\alpha \bigoplus_m (\bar{H}^{\otimes m})^{\oplus a_{i_\alpha,m}} \cong \bigoplus_m (\bar{H}^{\otimes m})^{\oplus b_m} ,$$

where b_m is the cardinal of a basis of the $\mathbb{F}_p$-vector space $\mathbb{F}_p^{\oplus a_{i_\alpha,m}}$.

Proof of 3.12.2. We treat the case $p > 2$. Define a $2\mathbb{N}[1/p]$-graded unstable $\mathcal{A}$-module L_*^* by

$$L_i^* = \begin{cases} \bar{H} & \text{if } i = 2p^j \text{ for some } j \in \mathbb{Z} \\ 0 & \text{otherwise} . \end{cases}$$

The Carlsson algebra K_*^* has a coproduct $\Delta : K_*^* \to K_*^* \otimes K_*^*$ which determines an $\mathbb{N} \times 2\mathbb{N}[1/p]$-graded Hopf algebra structure on K_*^*. The map Δ is completely determined by the formulas

$$\Delta \hat{x}_i = \hat{x}_i \otimes 1 + 1 \otimes \hat{x}_i ,$$
$$\Delta \hat{t}_i = \hat{t}_i \otimes 1 + 1 \otimes \hat{t}_i .$$

Moreover, the map Δ is clearly $\mathcal{A}$-linear.

Define a map g_* from the coaugmentation coideal $\bar{K}_*^*$ of K_*^* to L_*^* (recall that $\bar{K}_j^* = K_j^*$ if $j \neq 0$ and $\bar{K}_0^* = 0$) by:

— g_* preserves bidegree,

— $g_{2p^j} : K_{2p^j}^* \to L_{2p^j}^* = \bar{H}$ identifies with Carlsson's map g (see 3.1),

— $g_i = 0$ if $i \neq 2p^j$.

The map g_* lifts uniquely to an $\mathbb{N} \times 2\mathbb{N}[1/p]$-graded $\mathbb{F}_p$-coalgebra map $\hat{g}_*$ from K_*^* in the coalgebra cofreely cogenerated *in the non-commutative sense* by L_*^*. Denote it by $T(L_*^*)$, we have:

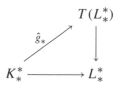

Remark. This situation is dual to the following, well known one; given a linear map from an $\mathbb{F}_p$-module M to an $\mathbb{F}_p$-algebra C, there is a unique multiplicative extension of this map to the tensor algebra $T(M)$ of M.

But $T(L_*^*)$, as an $(2\mathbb{N}[1/p]$-graded) unstable $\mathcal{A}$-module, is the direct sum $\oplus_{m \geq 0}(L_*^*)^{\otimes m}$, hence $T(L_*^*)_i \cong \oplus_{m \geq 0}(\bar{H}^{\otimes m})^{\oplus a_{i,m}}$ for certain integers $a_{i,m}$. The precise value of $a_{i,m}$ is the *finite* number of m-uples $(k_1, \ldots, k_m)$, $k_i \in \mathbb{Z}$, such that $2p^{k_1} + \ldots 2p^{k_m} = i$, but this precise value is not important there. Therefore, we have got an $\mathcal{A}$-linear map

$$\hat{g}_i : K_i^* \to \bigoplus_{m \geq 0} (\bar{H}^{\otimes m})^{\oplus a_{i,m}}.$$

It remains to prove that $\hat{g}_i$ is injective. This is a consequence of Proposition 3.9 of [**MM**] applied to the first grading. By the construction $\hat{g}_*$ is injective on primitive elements (these are the $\hat{t}_i$ and $\hat{x}_i^{p^n}$, $n \geq 0$, $i \in \mathbb{Z}$). Hence, as it is a coalgebra map, it is injective on K_*^*. Once again, the dual property for algebras is more familiar: a map of connected algebras which is surjective on indecomposable elements is surjective. This concludes the proof of (3.12.1).

H. Campbell and P. Selick [**CS**] have reproved the classification theorem in the following way. They have considered the algebra K_*^* quotiented out by the ideal generated by the elements $x_{i+n} - x_i$, for a given integer n and all integers i. This quotient is isomorphic, as an algebra, to $\mathbb{F}_2[x_0, \ldots, x_{n-1}]$. It is also provided with an unstable action of the Steenrod algebra for which the Cartan formula holds, and moreover satisfies $\text{Sq}^1 x_i = x_{i-1}^2$, $\text{Sq}^1 x_0 = x_{n-1}^2$. The remarkable fact is that as an $\mathcal{A}$-module it is isomorphic to $H^*(\mathbb{Z}/2^{\oplus n})$. From this one can recover the classification theorem.

3.13. The general case

Let I be an indecomposable $\mathcal{U}$-injective; we want to show

Theorem 3.13.1. *There exist an element $L \in \mathcal{L}$ and an integer n such that I is isomorphic to $L \otimes J(n)$.*

Lemma 3.13.2. *The unstable $\mathcal{A}$-module $L \otimes J(n)$ is indecomposable for any $L \in \mathcal{L}$, $n \in \mathbb{N}$.*

Lemma 3.13.3. *Let L and L' be in $\mathcal{L}$, n and n' in $\mathbb{N}$. If $L \otimes J(n)$ is isomorphic to $L' \otimes J(n')$, then $L = L'$ and $n = n'$.*

We shall need the following technical result.

Lemma 3.13.4. *Let I be an indecomposable $\mathcal{U}$-injective. Then there exists an integer n such that $\widetilde{\Sigma}^n I$ is trivial.*

Proof of (3.13.2).   We claim that $\widetilde{\Sigma}(L \otimes J(n))$ is isomorphic to $L \otimes J(n-1)$. This follows from (3.3.5) as L is a direct factor of $(H^*\mathbb{Z}/p)^{\otimes d}$ for some d and from the obvious isomorphism $\widetilde{\Sigma} J(n) \cong J(n-1)$. Therefore, if $L \otimes J(n)$ is isomorphic to $L' \oplus L''$, $L \otimes J(n-1)$ is isomorphic to $\widetilde{\Sigma} L' \oplus \widetilde{\Sigma} L''$. By induction, one can assume that $L \otimes J(n-1)$ is indecomposable. So, for example, $\widetilde{\Sigma} L'$ is trivial. But we claim that, if $n > 0$ and M is any non-zero sub $\mathcal{A}$-module $M \subset L \otimes J(n)$, $\widetilde{\Sigma} M$ is non-trivial.

To prove the claim filter $J(n)$ by degree. Consider the induced filtration on $L \otimes J(n)$ by taking the tensor product. The filtration is finite, the quotients are suspensions. Consider the induced filtration on M

$$M = M_0 \supset M_1 \supset \ldots \supset M_t \supset M_{t+1} = 0, \quad M_t \quad \text{non trivial},$$

clearly M_t is a non-zero suspension.

Therefore, L' is trivial and we are done.

Proof of (3.13.3).   One uses the same kind of argument as above, it is left to the reader.

Proof of (3.13.4).   The unstable $\mathcal{A}$-module $\Sigma^n \widetilde{\Sigma}^n I$ identifies with the largest n-th suspension contained in I. We have a decreasing filtration

$$I \supset \Sigma \widetilde{\Sigma} I \supset \Sigma^2 \widetilde{\Sigma}^2 I \supset \ldots \supset \Sigma^n \widetilde{\Sigma}^n I \supset \ldots .$$

Choose a non-zero element x in I. We claim that, for n large enough, $\mathcal{A}x \cap \Sigma^n \widetilde{\Sigma}^n I$ is trivial. This implies that $\Sigma^n \widetilde{\Sigma}^n I$ is trivial for n large enough because I is indecomposable (3.11.3). It remains to prove the claim. Let us start with $p = 2$.

We shall prove that, if $\mathcal{A}x \cap \Sigma^n \widetilde{\Sigma}^n I = \Sigma^n \widetilde{\Sigma}^n \mathcal{A}x$ is non-trivial for all n, there exists a sequence $(i_1, \ldots, i_k, \ldots)$ such that:

— for any n, $(i_n, \ldots, i_1)$ is admissible of excess less that $|x|$;

— $Sq^{i_{\alpha(n)}} \ldots Sq^{i_1} x \in \Sigma^n \widetilde{\Sigma}^n Ax$ and is non-zero, for a certain strictly increasing function $\alpha : \mathbb{N} \to \mathbb{N}$.

This leads to a contradiction because the excess of $(i_n, \ldots, i_1)$ must be bounded by $|x|$ and increases with n. Consequently it must be constant at a value e for n large enough; say if $n \geq m_0$. Therefore the sequence $(i_1, \ldots, i_m, \ldots)$ as the form $(i_1, \ldots, i_{m_0}, 2i_{m_0}, 4i_{m_0}, \ldots)$. But, if an element z of an unstable A-module belongs to an n-th suspension, one has $Sq^t z = 0$ as soon as $|z| - n < t$. For all $m > \sup(\alpha(n), m_0)$, the element $Sq^{2^{m-m_0} i_{m_0}} \ldots Sq^{i_{m_0}} Sq^{i_{m_0}-1} \ldots Sq^{i_1} x$ is in an n-th suspension. As it is non-zero, we must have

$$2^{m-m_0+1} i_{m_0} \leq |x| + i_1 + \ldots + i_{m_0-1} + i_{m_0}(2^{m-m_0+1} - 1) - n.$$

However, we may have chosen $n \gg |x|$ and under this hypothesis this inequality cannot hold.

We now prove the existence of the sequence $(i_1, \ldots, i_n, \ldots)$. Observe first that as $\widetilde{\Sigma}^n Ax$ is non-zero, for all n, there are infinitely many operations, and so infinitely many admissible monomials in the Sq^i's which are non-zero on x. If an admissible monomial $Sq^{j_t} \ldots Sq^{j_1}$ is non-zero on x, j_1 is less that $|x|$. Therefore, there exists a value i_1 for which there exist infinitely many admissible monomials of the form $Sq^{j_t} \ldots Sq^{j_2} Sq^{i_1}$ which are non-zero on x. The same argument applies: j_2 can take only finitely many values as it is less than $i_1 + |x|$. Therefore, there exists an integer i_2 for which there exist infinitely many admissible monomials of the form $Sq^{j_t} \ldots Sq^{j_3} Sq^{i_2} Sq^{i_1}$ which are non-zero on x. Continuing this way completes the proof of (3.13.4) when $p = 2$.

The case of an odd prime is treated similarly. We leave the details to the reader.

In order to complete the proof of Theorem 3.13.1, one proceeds as follows. Let I be an indecomposable $\mathcal{U}$-injective, let n be the largest integer for which $\widetilde{\Sigma}^n I$ is non-trivial (which exists by Lemma 3.13.4); $\widetilde{\Sigma}^n I$ is reduced and injective (since, if I is $\mathcal{U}$-injective, so is $\widetilde{\Sigma} I$). Moreover, it follows from (3.11.5) that $\widetilde{\Sigma}^n I$ is indecomposable. Therefore it is isomorphic to some $L \in \mathcal{L}$. By adjunction we get a map $\Sigma^n L \hookrightarrow I$ and by (3.11.5) I is the injective hull of $\Sigma^n L$. But, on the other hand, there is an obvious inclusion $\Sigma^n L \hookrightarrow L \otimes J(n)$, and by (3.13.2) and (3.11.5), $L \otimes J(n)$ is the injective hull of $\Sigma^n L$. Therefore, I is isomorphic to $L \otimes J(n)$ and we are done.

(3.13.5) Exercise [LS2]. Let M be a monogenic unstable $\mathcal{A}$-module. Assume that M is infinite. Show that there exist integers $i > 0$ and n such that $\mathrm{Hom}_{\mathcal{U}}(M, K(i) \otimes J(n))$ is non-trivial.

3.14. Applications

We list here some applications of the preceding sections.

Theorem 3.14.1. *Let I be a reduced injective unstable $\mathcal{A}$-module. Then there exists a unique family of cardinals (a_L), $L \in \mathcal{L}$ such that*

$$I \cong \bigoplus_{\mathcal{L}} L^{\oplus a_L} .$$

This is just a restatement of the main theorem in the reduced case, however it is worth stressing it.

Corollary 3.14.2. *Let I be a reduced injective unstable $\mathcal{A}$-module of finite type. Then there exists a unique family of integers $(a_L)_{L \in \mathcal{L}}$ such that*

$$I \cong \bigoplus_{\mathcal{L}} L^{\oplus a_L} .$$

Proof. This holds because in a given degree only finitely many members of $\mathcal{L}$ are non-trivial. In fact any L must be a direct factor of $\bar{H}^{\otimes q}$ for some q, and if L is non-zero in degree m, then $q \leq m$. But the number of direct summands which are non-trivial in degree m is bounded by the dimension of $(\bar{H}^{\otimes q})$ in degree m, which is finite. Note that this proof shows that a direct sum $\bigoplus_{\mathcal{L}} L^{\oplus a_L}$, where the a_L are integers, is of finite type.

Corollary 3.14.3. *Let M be an unstable $\mathcal{A}$-module. Then the following conditions are equivalent*

(i) *M is nilpotent;*

(ii) *for every elementary abelian p-group V, $\mathrm{Hom}_{\mathcal{U}}(M, H^*V)$ is trivial;*

(iii) *$T_V^0 M$ is trivial for every elementary abelian p-group V*

The implication (i) $\Rightarrow$ (ii) has been proved in Section 2.6. The conditions (ii) and (iii) are clearly equivalent. Thus it suffices to prove that (ii) implies (i).

Clearly this is a consequence of Proposition 3.12.2 and of the definition of the $K(i)$'s. It can also be seen as follows.

We first claim that the unstable $\mathcal{A}$-modules $L \otimes J(n)$, $n > 0$ are nilpotent. Indded consider the filtration $0 \subset J_n \subset J_{n-1} \subset \cdots \subset J_k \subset \cdots$ on $J(n)$ defined by $J_k^\ell = J(n)^\ell$, if $\ell \geq k$, $J_k^\ell = 0$ otherwise. This is a finite filtration by unstable $\mathcal{A}$-modules and the successive quotients are suspensions. The induced filtration on $L \otimes J(n)$ is again a finite filtration by unstable $\mathcal{A}$-modules and the successive quotients are suspensions. The claim follows because such a module is clearly nilpotent.

Let us come back to the proof of the implication. Let M be such that for all elementary abelian p-group V $\operatorname{Hom}_{\mathcal{U}}(M, H^*V)$ is trivial. There are no non-trivial map from M to a reduced injective unstable $\mathcal{A}$-module by Theorem 3.14.1. Thus its injective hull is a direct sum of $L \otimes J(n)$'s, $n > 0$. Thus it is a submodule of a nilpotent module. Hence it is nilpotent.

Theorem 3.14.4 (see [LS2]). *Let K be a reduced injective unstable $\mathcal{A}$-module and I be an injective unstable $\mathcal{A}$-module. Then the tensor product $K \otimes I$ is injective.*

This is a generalization of Corollary 2.8.2 and follows immediately from it and Theorem 3.11.1.

Theorem 3.14.5 (see [LS2]). *Let K be an unstable $\mathcal{A}$-module. Then the following conditions are equivalent:*

(i) *K is a reduced $\mathcal{U}$-injective;*

(ii) *for any $\mathcal{U}$-injective I, the tensor product $K \otimes I$ is still $\mathcal{U}$-injective;*

(iii) *$K \otimes J(d)$ is $\mathcal{U}$-injective for $0 \leq d \leq 2p - 2$.*

Only the implication (iii) $\Rightarrow$ (i) needs some explanation. It depends on Theorem 3.11.1 and on:

Lemma 3.14.6. *Let L be an element of $\mathcal{L}$ and n a strictly positive integer. There exists an integer d, with $1 \le d \le 2p - 2$ such that the injective dimension of $L \otimes J(n) \otimes J(d)$ is 1.*

Recall that the injective dimension is the minimal length of an injective resolution in the category $\mathcal{U}$.

The lemma is proved as follows. Assume first that $p = 2$ and that $n = 2k - 1$. The unstable $\mathcal{A}$-module $L \otimes J(2k - 1) \otimes J(1)$ has injective dimension 1. Indeed it has an injective resolution of the following form:

$$0 \to L \otimes (\Sigma J(2k - 1)) \to L \otimes J(2k) \xrightarrow{\pi} L \otimes J(k) \to 0,$$

where π denotes the tensor product of the identity map of L and of the map $\bullet \operatorname{Sq}^k$. This map has no section because $\widetilde{\Sigma}(\pi)$ is trivial and because the unstable $\mathcal{A}$-module $\widetilde{\Sigma}(L \otimes J(k))$ is not trivial.

In the case where $n = 2k$ consider the short exact sequence:

$$0 \to L \otimes J(2k) \otimes J(2) \to L \otimes (J(2k + 2) \oplus J(2k + 1))$$
$$\xrightarrow{\pi} L \otimes J(k + 1) \to 0,$$

where π denotes the tensor product of the identity map of L and of the map $\bullet \operatorname{Sq}^{k+1} \oplus \bullet \operatorname{Sq}^k : J(2k + 2) \oplus J(2k + 1) \to J(k + 1)$. Then one applies the functor $\widetilde{\Sigma}^2$, and one concludes using the same argument as above.

The case of an odd prime is left to the reader.

Note also the following where $p = 2$.

(3.14.7) Exercise (S. Zarati and the author). The unstable $\mathcal{A}$-module $J(n) \otimes J(m)$ is not injective as soon as $n \ge 2$ and $m \ge 2$.

(3.14.8) Exercise. State and prove a similar result for $p > 2$.

(3.14.9) Exercise. Alternative proof of the injectivity of $H^*\mathbb{Z}/2 \otimes J(n)$ (Lannes). The proof is done by induction on n by using the exact sequence:

$$0 \to H^*\mathbb{Z}/2 \otimes \Sigma J(n1) \to H^*\mathbb{Z}/2 \otimes J(n) \to H^*\mathbb{Z}/2 \otimes J(n/2) \to 0.$$

One applies the functor $\operatorname{Hom}_{\mathcal{U}}(M, -)$ and one shows that the group

$$\operatorname{Ext}^1_{\mathcal{U}}(M, H^*\mathbb{Z}/2 \otimes \Sigma J(n - 1))$$

is isomorphic to

$$\mathrm{Hom}_{\mathcal{U}}(\Omega_1 M,\, H^*\mathbb{Z}/2 \otimes J(n-1))\,.$$

Then one uses the formula $\Omega_1 M \cong \Sigma \Phi \widetilde{\Sigma} M$ (2.2.5) to identify it with the group

$$\mathrm{Hom}_{\mathcal{U}}(\widetilde{\Sigma} M,\, H^*\mathbb{Z}/2 \otimes \Sigma J((n-2)/2))\,.$$

Then one identifies the resulting map

$$\mathrm{Hom}_{\mathcal{U}}(M, H^*\mathbb{Z}/2 \otimes J(n/2)) \to \mathrm{Hom}_{\mathcal{U}}(\widetilde{\Sigma} M, H^*\mathbb{Z}/2 \otimes J((n-2)/2))$$

as the natural one. To do that one observes that any map

$$\varphi \circ \lambda_M :\, \Phi M \to M \to H^*\mathbb{Z}/2 \otimes J(n/2)$$

lifts to $H^*\mathbb{Z}/2 \otimes J(n)$ (compare with 2.3.6).

Deeper algebraic structure

4. The structure of indecomposable reduced $\mathcal{U}$-injectives

The aim of this chapter is to recollect the known facts about the structure (as graded vector spaces and $\mathcal{A}$-modules) of the indecomposable reduced $\mathcal{U}$-injectives. Unfortunately, because difficult problems in modular representation theory intervene, little is known — certainly much less than one would like to know.

4.1. The structure of the set $\mathcal{L}$

Let us define a subset $\mathcal{L}_d$, $d \geq 0$, of $\mathcal{L}$ (see 3.11) as follows: $L \in \mathcal{L}_d$ if and only if L embeds in $H^*(\mathbb{Z}/p^{\oplus d})$ but not in $H^*(\mathbb{Z}/p^{\oplus d-1})$. Clearly, $\mathcal{L}$ is the disjoint union of the $\mathcal{L}_d$, $d \geq 0$. From now on in this section, we shall denote $\mathbb{Z}/p^{\oplus d}$ by V. The map of algebras, from the semigroup ring $\mathbb{F}_p[\text{End}\,V]$ to the group ring $\mathbb{F}_p[\text{GL}\,V]$ which sends any singular element to zero and any regular element to itself, provides $\mathbb{F}_p[\text{GL}\,V]$ with an $\mathbb{F}_p[\text{End}\,V]$-bimodule structure.

For any unstable $\mathcal{A}$-module M, allowing $\text{End}\,V$ to act on H^*V in the natural way, the vector space $\text{Hom}_{\mathcal{U}}(M, H^*V)$ is a right $\mathbb{F}_p[\text{End}\,V]$-module. Considering $\mathbb{F}_p[\text{GL}\,V]$ as a left $\mathbb{F}_p[\text{End}\,V]$-module, we define $R_d(M)$ by the formula

$$\text{Hom}_{\mathcal{U}}(M, H^*V) \bigotimes_{\mathbb{F}_p[\text{End}\,V]} \mathbb{F}_p[\text{GL}\,V].$$

By the Adams-Gunawardena-Miller, Lannes theorem, $R_d(H^*V)$ is isomorphic to $\mathbb{F}_p[\text{GL}\,V]$ (as a right $\mathbb{F}_p[\text{GL}\,V]$-module). Therefore, it is clear that if L is an element of $\coprod_0^d \mathcal{L}_i$, hence a direct summand of H^*V, $R_d(L)$ is a direct summand of $\mathbb{F}_p[\mathrm{GL}\,V]$. Consequently, $R_d(L)$ is a projective (right) $\mathbb{F}_p[\mathrm{GL}\,V]$-module.

Let $\mathcal{P}_d$ be a set of representatives of indecomposable projective $\mathbb{F}_p[\mathrm{GL}\,V]$-modules. The following has been proved by Harris and Kuhn [HK]:

Theorem 4.1.1. *The map* R_d *determines a bijection between* $\mathcal{L}_d$ *and* $\mathcal{P}_d$. *Therefore* $\mathcal{L}_d$ *has cardinality* $p^d - p^{d-1}$.

Harris and Kuhn give two proofs of this result, one depending on the representation theory of the semigroup ring $\mathbb{F}_p[\mathrm{End}\,V]$. The other one is close from an idea of Nishida [Ni]. We are going to present an algebraic version of this last one (see also [HS]). It is a consequence of the following lemma that shows that R_d determines a map from $\mathcal{L}_d$ to $\mathcal{P}_d$.

Lemma 4.1.2.

a) *If* $L \in \mathcal{L}_i$, $i < d$, *then* $R_d(L)$ *is trivial;*

b) *if* $L \in \mathcal{L}_d$, *then* $R_d(L)$ *is non-trivial,*

c) *moreover it is an indecomposable projective* $\mathbb{F}_p[\mathrm{GL}\,V]$-*module.*

Proof.

Part a) This is almost by definition and is left to the reader (use the Adams-Gunawardena-Miller, Lannes theorem, i.e. (3.9.1) in degree zero to show that $R_d(H^*(\mathbb{Z}/p^{\oplus d-1})) = 0$).

Part b) We show that $R_d(L)$ is non-trivial. Consider the exact sequence of left $\mathbb{F}_p[\mathrm{End}\,V]$-modules:

$$\bigoplus_{0 \neq W \subsetneq V} \mathbb{F}_p[\mathrm{Hom}(V/W, V)] \rightarrow \mathbb{F}_p[\mathrm{End}\,V] \rightarrow \mathbb{F}_p[\mathrm{GL}\,V] \rightarrow 0,$$

and apply to it the functor $S \mapsto \mathrm{Hom}_{\mathcal{U}}(L, H^*V) \otimes_{\mathbb{F}_p[\mathrm{End}\,V]} S$. As $\mathrm{Hom}_{\mathcal{U}}(L, H^*V)$ is projective we get an exact sequence of right $\mathbb{F}_p[\mathrm{End}\,V]$-modules. If $R_d(L)$ is trivial, one deduces from this exact sequence that any embedding $e : L \hookrightarrow H^*V$ can be written as a sum $\sum_k \alpha_k^* \circ f_k$, where $\alpha_k : V \rightarrow V$ has a non trivial kernel W_k, and

$f_k \in \mathrm{Hom}_{\mathcal{U}}(L, H^*V)$. Let $\tilde{\alpha}_k : V/W_k \to V$ be the map induced by α_k. Then the map

$$\bigoplus_k \tilde{\alpha}_k^* \circ f_k : L \to \bigoplus_k H^*(V/W_k)$$

is an embedding. As L is indecomposable, it follows from (3.11.6) that one of the maps $\tilde{\alpha}_k^* \circ f_k$ is injective. Therefore, L belongs to $\mathcal{L}_i$ for $i = \dim V/W_k$; this is a contradiction as $i < d$.

Part c) As the ring $\mathrm{End}_{\mathcal{U}}(H^*V)$ is isomorphic to the semigroup ring $\mathbb{F}_p[\mathrm{End}\,V]$, the map $I \mapsto \mathrm{Hom}_{\mathcal{U}}(I, H^*V)$ determines a bijection, from a set of representatives of indecomposable direct summands of H^*V, to a set of representatives of indecomposable direct summands of $\mathbb{F}_p[\mathrm{End}\,V]$ (projective indecomposable $\mathbb{F}_p[\mathrm{End}\,V]$-modules). The inverse map is determined by the assignment that associates to a projective indecomposable P 'the unstable $\mathcal{A}$-module' $\mathrm{Hom}_{\mathbb{F}_p[\mathrm{End}\,V]}(P, H^*V)$. Let I be in $\mathcal{L}_d$. The $\mathbb{F}_p[\mathrm{End}\,V]$-module, $\mathrm{Hom}_{\mathcal{U}}(I, H^*V)$ is indecomposable. As we have an $\mathbb{F}_p[\mathrm{End}\,V]$-linear surjection

$$\mathrm{Hom}_{\mathcal{U}}(I, H^*V) \to R_d(I),$$

and as $R_d(I)$ is non-trivial $\mathrm{Hom}_{\mathcal{U}}(I, H^*V)$ is the projective cover, as $\mathbb{F}_p[\mathrm{End}\,V]$-module, of $R_d(I)$ **[CR]** ($R_d(I)$ is considered as an $\mathbb{F}_p[\mathrm{End}\,V]$-module via the above surjection $\mathbb{F}_p[\mathrm{End}\,V] \to \mathbb{F}_p[\mathrm{GL}\,V]$). If $R_d(I)$ is decomposable as an $\mathbb{F}_p[\mathrm{GL}\,V]$-module, it is also decomposable as an $\mathbb{F}_p[\mathrm{End}\,V]$-module and its projective cover is decomposable. This is a contradiction.

The theorem follows. Indeed, R_d is clearly surjective, it is injective because $R_d(I)$ determines its projective cover as an $\mathbb{F}_p[\mathrm{End}\,V]$-module. Thus $R_d(I)$ determines I.

4.2. Results on the Poincaré series of indecomposable reduced $\mathcal{U}$-injectives

It would be important to know more precisely the structure of the indecomposable, reduced $\mathcal{U}$-injectives. First of all one would like to know their structure as graded vector space, i.e. their Poincaré series. Unfortunately, very little is known about these series. The aim of this section is to collect some of the known results, mostly without proofs. As some the following results are only true at the prime 2, we shall only consider this case and add a comment when they extend to $p > 2$.

If ρ is an element of $\mathcal{P}_d$, let L_ρ be the corresponding element of $\mathcal{L}_d$ and $P_{L_\rho}(q)$ its Poincaré series.

Theorem 4.2.1 [Mt2]. *The Poincaré series $P_{L_\rho}(q)$ is a rational function which has a pole of order d at $q = 1$.*

This extends to $p > 2$.

In order to state the other results, we need to introduce some notations. The (isomorphism classes of) simple $\mathbb{F}_2$-representations of $GL_d \, \mathbb{F}_2$ are indexed by partitions $(\lambda_1, \ldots, \lambda_d)$ such that

$$\lambda_1 \geq \ldots \geq \lambda_d = 1$$
$$\lambda_i - \lambda_{i+1} \leq 1 \, , \quad i \leq d - 1 \, .$$

Such partitions are called column 2-regular. We refer to **[J]** and **[JK]** for more information. One checks that there are $2^d - 2^{d-1}$ such partitions (this is the number of 2-regular conjugacy classes in $GL_d \, \mathbb{F}_2$).

Recall that the process of taking the projective cover determines a bijection from the set of isomorphism classes of simple $\mathbb{F}_2$-representations of $GL_d \, \mathbb{F}_2$ to $\mathcal{P}_d$ (the inverse map is determined by the 'socle', *i.e* the uniquely defined simple subobject).

We denote by I_λ be the simple representation of $GL_d \, \mathbb{F}_2$ associated to be the column 2-regular partition λ, by P_λ the projective cover, and let L_{P_λ} (L_λ for short) be the corresponding element of $\mathcal{L}_d$.

Examples.
If $d = 2$, $GL_2 \, \mathbb{F}_2 = \mathfrak{S}_3$, and there are two simple representations:
(i) the trivial one, $\mathbb{F}_2$, which corresponds to $(1, 1)$ and which has projective cover $\mathbb{F}_2 [GL_2 \, \mathbb{F}_2 / \mathbb{Z}/3]$,
(ii) the standard one $\mathbb{Z}/2^{\oplus 2}$ which is its own projective cover and corresponds to $(2,1)$.

If $d = 3$, there are four simple representations:

(i)   the trivial one, $\mathbb{F}_2$, which corresponds to $(1, 1, 1)$,

(ii) the natural one, $V = \mathbb{F}_2^{\oplus 3}$, which corresponds to $(2, 1, 1)$,

(iii) the contragredient, $V^* = \Lambda^2 V$, which corresponds to $(2, 2, 1)$,

(iv) the Steinberg representation which is of dimension 8 and corresponds to $(3, 2, 1)$.

For a description of the projective covers, when $d = 3$ we refer to **[Mt1]**.

Of special importance is the case of the Steinberg representation. It corresponds to the partition $\Delta_d = (d, d - 1, \ldots, 1)$. This is the only (up to isomorphism) simple representation of $\mathrm{GL}_d\, \mathbb{F}_2$ which is projective. It belongs to the block of defect zero of $\mathbb{F}_2\,[\mathrm{GL}_d\, \mathbb{F}_2\,]$ (and is -up to isomorphism- the only one to belong to the block of defect zero) **[CR]**. The modular character of the Steinberg representation, St_d, is known. There is also an explicit formula for an idempotent $e_d \in \mathbb{F}_2\,[\mathrm{GL}_d\, \mathbb{F}_2\,]$, such that the representation St_d is isomorphic to $e_d \mathbb{F}_2\,[\mathrm{GL}_d\, \mathbb{F}_2\,]$.

Theorem 4.2.2 [MtP]. *The Poincaré series $P_{L_{\Delta_d}}(q)$ is given by the formula*

$$\frac{q^{1+3+\ldots+2^d-1}}{(1 - q)(1 - q^3)\ldots(1 - q^{2^d-1})}.$$

The analogue is known for $p > 2$ (see **[MtP]**, Remark 5.13).

The original proof of S. Mitchell and S. Priddy depends on the following topological result. Denote by $M(d)$ the spectrum $e_d B\mathbb{Z}/2^{\oplus d}$. It is the telescope of the stable map from $B\mathbb{Z}/2^{\oplus d}$ to itself induced by e_d. By the very definition of $M(d)$, the cohomology $H^* M(d)$ is isomorphic to $e_d H^*(\mathbb{Z}/2^{\oplus d}) \cong \mathrm{Hom}_{\mathbb{F}_2[\mathrm{GL}_d\, \mathbb{F}_2\,]}(\mathrm{St}_d, H^*(\mathbb{Z}/2^{\oplus d}))$. This reduced $\mathcal{U}$-injective is not indecomposable. Mitchell and Priddy prove that the spectrum $M(d)$ splits as a wedge $L(d) \vee L(d - 1)$ of indecomposable spectra such that $H^* L(d) \cong L_{\Delta_d}$ (see also Proposition 4.2.5). Then

Theorem 4.2.3 [MtP]. *The spectrum $L(d)$ is homotopy equivalent to the cofiber of the inclusion $\Sigma^{-d} \mathrm{Sp}^{p^{d-1}} S^0 \hookrightarrow \Sigma^{-d} \mathrm{Sp}^{p^d} S^0$.*

Here $\mathrm{Sp}^k S^0$ is the k-symmetric product spectrum over S^0. But the spectrum $\mathrm{Sp}^\infty S^0$ is homotopically equivalent to $K(\mathbb{Z}, 0)$ by the Dold-Thom theorem. Then Nakaoka's theorem [Na] says that a basis for $H^* \mathrm{Sp}^{p^d} S^0$ is given by those admissible monomials $\mathrm{Sq}^I = \mathrm{Sq}^{i_1} \ldots \mathrm{Sq}^{i_r}$ such that $r \le d$ and $i_r \ge 2$. This allows computation of the Poincaré series of the theorem. Moreover this determines the $\mathcal{A}$-module structure of $H^* M(d)$. Indeed, $H^* \mathrm{Sp}^d S^0$ is isomorphic (as $\mathcal{A}$-module) to the quotient of $\mathcal{A}$ by the sub $\mathcal{A}$-module with $\mathbb{F}_2$-basis the admissible monomials $\mathrm{Sq}^I = \mathrm{Sq}^{i_1} \ldots \mathrm{Sq}^{i_r}$ with $i_r \ge 2$, $r > d$ and $\mathrm{Sq}^I \mathrm{Sq}^1$, all I.

In their proof, Mitchell and Priddy show that $H^* M(d)$ is isomorphic to the intersection of the span over $\mathcal{A}$ of the class $(x_1 \ldots x_d)^{-1} \in \mathbb{F}_2[x_1, \ldots, x_d, x_1^{-1}, \ldots, x_d^{-1}]$ with $\mathbb{F}_2[x_1 \ldots x_d] = H^*(\mathbb{Z}/2^{\oplus d})$ (see 3.5 and 5.6 in [MtP]). This intersection is spanned by the classes $\mathrm{Sq}^I (x_1 \ldots x_d)^{-1}$, $\ell(I) = d$, and $H^* M(d-1)$ corresponds to classes with I ending in 1.

Later a 'purely algebraic' computation of the Poincaré series was found by N. Kuhn and S. Mitchell [KM].

D. Carlisle and G. Walker have extended 4.2.2 to "the representations of $\mathrm{GL}(n, p)$ which are the closest neighbours of St in the $\mathrm{SL}(n, p)$ weight diagram" (see [CW]).

Beyond the preceding results, the only known general fact concerns the first coefficient of the Poincaré series of an indecomposable $\mathcal{U}$-injective.

Theorem 4.2.4. *Let* $\lambda = (\lambda_1, \ldots, \lambda_d)$ *be a column 2-regular partition (i.e.* $1 \ge \lambda_i - \lambda_{i+1} \ge 0$*) such that,* $\lambda_d = 1$*. Then:*

$$P_{L_\lambda}(q) = q^{\lambda_1' + 2\lambda_2' + 4\lambda_4' + \ldots + 2^{t-1}\lambda_t'} + \text{terms of higher degree} .$$

Here $(\lambda_1', \ldots, \lambda_t')$ *is the conjugate partition of* $\lambda : \lambda_i' = \#\{j / i \le \lambda_j\}$*.*

This was first proved by the author [FS] using 'mainly' the Steenrod algebra, then D. Carlisle and N. Kuhn gave a 'purely algebraic' proof in [CK].

Here are the examples which have been completely computed; we denote here $P_{L_\lambda}(q)$ by $P_\lambda(q)$:

If $d = 2$,

$$P_{(1,1)}(q) = \frac{q^2 + q^3}{(1-q)(1-q^3)}$$

$$P_{(2,1)}(q) = \frac{q^4}{(1-q)(1-q^3)} \quad \text{(is given by 4.2.2)}.$$

If $d = 3$, the complete computation has been done by S. Mitchell in [**Mt1**]. If $d = 4$, the computation has been done by D. Carlisle.

At odd primes the situation is completely known up to $d = 3$. This last case is due to S. Doty and G. Walker [**DtW**].

In order to finish this section, we want to explain with more details our claim that the knowledge of $P_{L_\lambda}(q)$ depends on results on modular representation theory, and give the proof of Theorem 4.2.1. Let λ be a partition (satisfying the requirements above). Denote by $\bar\lambda$ the partition $(\lambda_1 - 1, \ldots, \lambda_d - 1)$. Associated to $\bar\lambda$ there is a simple representation of some $GL_{d'}\, \mathbb{F}_2$, $d' < d$. The following has been proved by Harris and Kuhn [**HK**]:

Proposition 4.2.5. *The* $\mathcal{U}$-*injective*

$$E_\lambda = \mathrm{Hom}_{\mathbb{F}_2[GL_d\, \mathbb{F}_2]}(P_\lambda, H^*(\mathbb{Z}/2^{\oplus d}))$$

is isomorphic to $L_\lambda \oplus L_{\bar\lambda}$.

We refer to [**HK**] for the proof and for the case $p > 2$. It is easy to prove the following weaker result: E_λ is isomorphic to the direct sum of L_λ and of elements of $\coprod_0^{d-1} \mathcal{L}_i$. The interest of Proposition 4.2.5 is that, if one knows $P_{E_\lambda}(q)$ for all λ, one can deduce, by induction, $P_{L_\lambda}(q)$ for all λ. And, as we are going to explain, the computation of $P_{E_\lambda}(q)$ reduces to a (difficult) problem of representation theory. Recall that we denote by P_λ the projective cover (in the category of $\mathbb{F}_2[GL_d\, \mathbb{F}_2]$-modules) of the simple $\mathbb{F}_2[GL_d\, \mathbb{F}_2]$-module I_λ.

The series $P_{E_\lambda}(q)$ is described as follows. For an $\mathbb{F}_2[GL_d\, \mathbb{F}_2]$-module M, denote by $\langle I_\lambda, M \rangle = \dim_{\mathbb{F}_2} \mathrm{Hom}_{\mathbb{F}_2[GL_d\, \mathbb{F}_2]}(P_\lambda, M)$ the number of occurrences of I_λ in the Jordan-Hölder sequence of M. Then

$$P_{E_\lambda}(q) = \sum_{n \geq 0} \Big\langle I_\lambda, H^n(\mathbb{Z}/2^{\oplus d}) \Big\rangle q^n.$$

This can be expressed in terms of modular characters (see [CR] or [S2]). As the modular character of $H^n(\mathbb{Z}/2^{\oplus d})$ is known, it would be 'enough' to know the modular character of P_λ to do the computation (in the case $d = 4$ one can use [Be]).

Consider the subalgebra $(H^*(\mathbb{Z}/2^{\oplus d}))^{\mathrm{GL}_d}\,\mathbb{F}_2$ of Dickson invariants. This is a polynomial algebra with generators in the degrees $2^d - 2^{d-1}$, $2^d - 2^{d-2}, \ldots, 2^d - 1$. Denote it by $D^*(d)$. The algebra $H^*(\mathbb{Z}/2^{\oplus d})$ is a free $D^*(d)$-module with $(2^d - 2^{d-1})\ldots(2^d - 1)$ generators. Consider the *finite* graded $\mathbb{F}_2[\mathrm{GL}_d\,\mathbb{F}_2]$-module $M = H^*(\mathbb{Z}/2^{\oplus d})\otimes_{D^*(d)}\mathbb{F}_2$. The $\mathbb{F}_2[\mathrm{GL}_d\,\mathbb{F}_2]$-modules $H^*(\mathbb{Z}/2^{\oplus d})$ and $M\otimes D^*(d)$ have the same Jordan-Hölder series in each degree. Therefore

$$P_{E_\lambda}(q) = \sum_{n\geq 0}\langle I_\lambda, M^n\rangle q^n/(1 - q^{2^d - 2^{d-1}})\ldots(1 - q^{2^d - 1}),$$

the numerator being a polynomial with non-negative coefficients. Theorem 4.2.1 follows. There is another proof in [CW] that yields the better denominator $(1 - q^{2^d - 1})\ldots(1 - q^{2 - 1})$.

4.3. The decomposition of the Carlsson modules $K(i)$

As in the preceding section, we work at the prime 2, but the results extend without problems to $p > 2$. From Chapter 3, we know there exist integers $(a_{i,L})_{L\in\mathcal{L}}$ such that $K(i)$ is isomorphic to the direct sum $\bigoplus_{\mathcal{L}} L^{\oplus a_{i,L}}$. These integers $a_{i,L}$ can be computed in terms of modular representation theory invariants. We give the formula established by Lannes and the author.

Recall a few facts about the modular representation theory of the symmetric groups $\mathfrak{S}_n$ [J][JK]. The simple $\mathbb{F}_2$-representations are (up to isomorphism) indexed by partitions $\lambda = (\lambda_1, \ldots, \lambda_d)$ of n such that $\lambda_i > \lambda_{i+1}$, $1 \leq i \leq d - 1$. Such partitions are called 2-regular. Observe that the conjugate partition (see Theorem 4.2.4) λ' satisfies the requirement of (4.2) ($1 \geq \lambda'_i - \lambda'_{i+1} \geq 0$), i.e. is column 2-regular. Note that $(\lambda')' = \lambda$. Let L_λ be in $\mathcal{L}_d$, let $V_{\lambda'}$ be the simple $\mathbb{F}_2$-representation of $\mathfrak{S}_k$ indexed by λ', where $k = \Sigma\lambda'_i = \Sigma\lambda_j$. Then

(4.3.1) $a_{i,L_\lambda} = \sum_\mu s(i,\mu)\dim_{\mathbb{F}_2}\bar{\mathfrak{S}}_\mu V_{\lambda'}.$

In this formula

— $\mu = (\mu_1, \ldots, \mu_\ell, \ldots, \mu_t)$ runs through the set of all partitions of k,

— $s(i, \mu)$ is the number of ways to write i as $\sum_\ell \mu_\ell 2^{q_\ell}$, $q_\ell \in \mathbb{Z}$ and $\ell \neq h \Rightarrow q_\ell \neq q_h$,

— $\mathfrak{S}_\mu \subset \mathfrak{S}_k$ is a Young subgroup isomorphic to $\mathfrak{S}_{\mu_1} \times \ldots \times \mathfrak{S}_{\mu_t}$, $\bar{\mathfrak{S}}_\mu = \sum_{\sigma \in \mathfrak{S}_\mu} \sigma \in \mathbb{F}_2[\mathfrak{S}_k]$ and $\bar{\mathfrak{S}}_\mu V_{\lambda'}$ is the image in $V_{\lambda'}$ of the norm with respect to $\mathfrak{S}_\mu$ (multiplication by $\bar{\mathfrak{S}}_\mu$).

Examples.

(i)   Denote by (1^n) the partition $(1, \ldots, 1)$ where 1 occurs n times. $L_{(1^n)}$ is not a direct summand in $K(1)$ if $n > 1$. The corresponding representation of $\mathfrak{S}_n$ is the trivial one. Therefore, $\dim_{\mathbb{F}_2} \bar{\mathfrak{S}}_\mu \mathbb{F}_2$ is zero unless $\mu = (1, \ldots, 1)$. But, in this case $s(1, \mu)$ is zero.

(ii) $L_{(2,1)}$ occurs once in $K(3)$, for the only partition of 3 for which $s(3, \mu)$ is non zero is $(2, 1)$. Moreover the representation of $\mathfrak{S}_3$ indexed by $(2, 1)$ is $\mathbb{F}_2^{\oplus 2}$ with the natural action of $\mathrm{GL}_2 \, \mathbb{F}_2 \cong \mathfrak{S}_3$ since $s(3, (2, 1)) = 1$.

(iii) $L_{(2,1)}$ occurs twice in $K(5)$. This is the same analysis as above, but $s(5, (2, 1)) = 2$.

(iv) $L_{(n,n-1,\ldots,1)}$ occurs in $K(i)$ if and only if $\alpha(i) \leq n(n+1)/2$.

We refer to Chapter 5 for the proof of formula 4.3.1.

4.4. Information about the $\mathcal{A}$-module structure of the reduced indecomposable $\mathcal{U}$-injectives

Outside of the case of the Steinberg module of Mitchell and Priddy, very little is known about the $\mathcal{A}$-module structure of the reduced indecomposable $\mathcal{U}$-injectives. We record here two results. The first one is due to Carlisle and Kuhn.

Let $\mathcal{A}(n)$ be the subalgebra of the mod 2 Steenrod algebra generated by $\mathrm{Sq}^1, \ldots, \mathrm{Sq}^{2^i}, \ldots, i \leq n - 1$. Then:

Theorem 4.4.1. *The indecomposable reduced $\mathcal{U}$-injective E_λ is free as an $\mathcal{A}(n)$-module if and only if $\lambda_1 \leq n$.*

We refer to [**CK1**] for the proof.

The second result, due to the author, describes the smallest non-trivial $\mathcal{N}il$-closed module contained in E_λ. For a precise definition of $\mathcal{N}il$-closed module, we refer to Chapter 6. For the moment, let us just say that a sub $\mathcal{A}$-module N of E_λ is $\mathcal{N}il$-closed if and only if the following holds:

$$(y \in E_\lambda \text{ and } \mathrm{Sq}_0 \, y \in N) \;\Rightarrow\; y \in N \, .$$

The fact that there exists a smallest non-trivial sub $\mathcal{A}$-module in E_λ which is $\mathcal{N}il$-closed follows from the generalities about indecomposable injective modules.

Denote by n the sum of the λ_i. Let (as in 4.3) $V_{\lambda'}$ be the simple $\mathbb{F}_2$-representation of $\mathfrak{S}_n$ indexed by λ'. Let $e_{\lambda'}$ be an element of $\mathbb{F}_2[\mathfrak{S}_n]$ such that $V_{\lambda'} \cong e_{\lambda'}\mathbb{F}_2[\mathfrak{S}_n]$; such an element exists by the general theory of modular representations. A formula for $e_{\lambda'}$ is given in [**J**], (see also [**JK**]). Denote by S_λ the smallest non-trivial $\mathcal{N}il$-closed sub $\mathcal{A}$-module of E_λ.

Theorem 4.4.2.

a) *The unstable $\mathcal{A}$-module S_λ is isomorphic to $e_{\lambda'}F(1)^{\otimes n}$ (where we allow $\mathfrak{S}_n$ to act on $F(1)^{\otimes n}$ by permutation of the factors);*

b) *The unstable $\mathcal{A}$-module S_λ is cyclic (over $\mathcal{A}$).*

One reason to justify the interest of S_λ is that in the category $\mathcal{U}/\mathcal{N}il$ (to be defined in Chapter 5), S_λ is the socle of E_λ. The S_λ's are the simple objects of the category $\mathcal{U}/\mathcal{N}il$.

Most of the arguments necessary to the proof of (4.4.2. a)) are in [**FS**]. Then (4.4.2. b) comes as a generalization of [**Mt3**] (the argument has been written down in details by P. Krason in his thesis).

We note to conclude that it is possible to show the dimension of the bottom class of S_λ is $\lambda_1 + 2\lambda_2 + \dots$.

5. The category $\mathcal{U}/\mathcal{N}il$, analytic functors, and representations of the symmetric groups

In this chapter we study the category $\mathcal{U}$ quotiented by its full subcategory consisting of nilpotent objects which is denoted by $\mathcal{N}il$. As we shall see in Chapter 6 this category is a building block for $\mathcal{U}$. We shall interpret it as a certain category of functors (see Theorem 5.2.6). Then, in Section 5.5, we define a filtration on $\mathcal{U}/\mathcal{N}il$ and we identify the quotient categories of the filtration with the categories of modular representations of the symmetric groups (Theorem 5.5.1). We explain in the same section the links with the preceding description. Then we describe simple objects in this category and prove, as a corollary, Formula 4.3.1.

5.1. The category $\mathcal{U}/\mathcal{N}il$ and the functor $\bar{f}$

Recall the notion of quotient category. Let $\mathcal{C}$ be an abelian category, $\mathcal{D}$ a full subcategory which has the property that, if we are given a short exact sequence in $\mathcal{C}$

$$0 \to C' \to C \to C'' \to 0,$$

if any two objects are in $\mathcal{D}$, so is the third. We shall say that $\mathcal{D}$ is a Serre class (in $\mathcal{C}$). This condition is a bit stronger than the one used in [Gb]. With this assumption, it is possible to define the quotient category $\mathcal{C}/\mathcal{D}$ as follows:

— the objects are the same as in $\mathcal{C}$;

— if we are given $S, T \in \mathcal{C}$, $\text{Hom}_{\mathcal{C}/\mathcal{D}}(S, T)$ is defined to be

$$\text{colim}_{(S', T')} \text{Hom}_{\mathcal{C}}(S', T/T'),$$

where (S', T') runs through the set of subobjects S' of S, T' of T such that $S/S' \in \mathcal{D}$ and $T' \in \mathcal{D}$.

Roughly speaking, through this process a morphism (in $\mathcal{C}$) $\varphi : S \to T$ such that $\text{Ker } \varphi \in \mathcal{D}$ (resp. $\text{Coker } \varphi \in \mathcal{D}$) becomes a monomorphism (resp. an epimorphism) in $\mathcal{C}/\mathcal{D}$. We refer to [Gb] for more details about the construction of $\mathcal{C}/\mathcal{D}$.

There is a forgetful functor $\mathcal{C} \to \mathcal{C}/\mathcal{D}$ which is exact and has the following universal property. If one is given an exact functor F from $\mathcal{C}$ to an abelian category $\mathcal{F}$ and if F is trivial on $\mathcal{D}$, there is a unique factorization

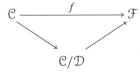

Denote by $\mathcal{N}il$ the full subcategory of $\mathcal{U}$ consisting of nilpotent objects. The category $\mathcal{N}il$ is clearly a Serre class in $\mathcal{U}$. Thus we can consider the quotient category $\mathcal{U}/\mathcal{N}il$. The forgetful functor $\mathcal{U} \to \mathcal{U}/\mathcal{N}il$ is going to be denoted by r_1 (see Chapter 6).

In some sense, looking at $\mathcal{U}/\mathcal{N}il$, is restricting attention to $\mathcal{N}il$-closed modules in $\mathcal{U}$. Indeed the forgetful functor $\mathcal{U} \to \mathcal{U}/\mathcal{N}il$ has a right adjoint s_1 (see Corollary 2.2.2 and Chapter 6). It follows from the adjunction property that the unstable module $s_1 r_1(M)$ is reduced. It is, in some sense for $p = 2$, 'quadratically closed' (again see Chapter 6) and the unit of the adjunction $M \to s_1 r_1(M)$ is initial for maps into reduced and 'quadratically closed' (for $p = 2$) unstable modules. The modules which satisfy these conditions are also called Δ-closed (see [AW] and [SS]).

In the case of an odd prime we leave it as an exercise to the reader to check that there is an equivalence of categories:

$$\mathcal{U}/\mathcal{N}il \cong \mathcal{U}'/\mathcal{N}il',$$

the category $\mathcal{U}'$ (resp. $\mathcal{N}il'$) being the category of evenly graded (resp. nilpotent) unstable $\mathcal{A}$-modules. This allows to work in $\mathcal{U}'$.

Denote by $\mathcal{F}$ the category of covariant functors from the category $\mathcal{E}_f$ of finite dimensional $\mathbb{F}_p$-vector spaces to the category $\mathcal{E}$. It is an abelian category. A short exact sequence of functors is a sequence $0 \to F' \to F \to F'' \to 0$, such that for any V,

$$0 \to F'(V) \to F(V) \to F''(V) \to 0 .$$

is exact.

Following [HLS1], let us introduce a functor $\bar{f} : \mathcal{U} \to \mathcal{F}$. This functor associates to an object M of $\mathcal{U}$ the functor

$$\mathcal{E}_f \to \mathcal{E}, \bar{f}(M) : V \longmapsto T_V(M)^0 .$$

The exactness of the functor T_V (in degree 0) immediately implies

Proposition 5.1.1. *The functor $\bar{f}$ is exact.*

The category $\mathcal{F}$ has a tensor product. The tensor product of the functors F and G is the functor $F \otimes G$ that associates to V the $\mathbb{F}_p$-vector space $F(V) \otimes G(V)$. A direct consequence of the natural isomorphism $T_V(M) \otimes T_V(N) \cong T_V(M \otimes N)$ is:

Proposition 5.1.2. *For any unstable A-modules M and N, there is a natural equivalence between the functors $\bar{f}(M) \otimes \bar{f}(N)$ and $\bar{f}(M \otimes N)$.*

The category $\mathcal{F}$ has colimits, in fact

$$(\mathrm{colim}_\alpha F_\alpha)(V) = \mathrm{colim}_\alpha F_\alpha(V) .$$

Proposition 5.1.3. *The functor $\bar{f}$ commutes with colimits.*

(5.1.4) Examples. If M is the unstable A-module $F(n)$, then by the very definition, $\bar{f}\big(F(n)\big)(V)$ is equivalent to $H_n(V)$, the n-th mod p homology group of V.

If M is $F(1)^{\otimes n}$, using (5.1.2) we get $\bar{f}\big(F(1)^{\otimes n}\big)(V) \cong V^{\otimes n}$.

The next example is fundamental. It is a consequence either of the Adams-Gunawardena-Miller theorem, or of Lannes' linearization principle. Let M be H^*W then $\bar{f}(H^*W)(V)$ is equivalent to $\mathbb{F}_p^{\mathrm{Hom}(V,W)}$, the space of set maps from $\mathrm{Hom}(V, W)$ to $\mathbb{F}_p$.

Our aim is to describe the 'kernel' of $\bar{f}$ and its image in $\mathcal{F}$. We begin with the kernel. If M is a nilpotent unstable $\mathcal{A}$-module $T_V(M)^0$ is trivial for any V. This is the easy implication of Corollary 3.14.3. As the functor $\bar{f}$ is exact, the fundamental property of the quotient category $\mathcal{U}/\mathcal{N}il$ implies that $\bar{f}$ factors (uniquely up to natural equivalence) as

$$\mathcal{U} \xrightarrow{r} \mathcal{U}/\mathcal{N}il \xrightarrow{f} \mathcal{F},$$

where r is the canonical functor. This factorization defines the functor f. The difficult implication of Corollary 3.14.3. can be restated as

Proposition 5.1.5. *The functor* $f : \mathcal{U}/\mathcal{N}il \to \mathcal{F}$ *is faithful.*

We are left with the task of describing the image of f in $\mathcal{F}$.

5.2. Analytic functor

Let F be an object of $\mathcal{F}$, define ΔF to be the functor that associates to V the kernel of the linear map $F(\pi) : F(V \oplus \mathbb{F}_p) \to F(V)$, where $\pi : V \oplus \mathbb{F}_p \to V$ is the standard projection: the identity on V and zero on the summand $\mathbb{F}_p$. It could be as well defined as the cokernel of $F(i) : F(V) \to F(V \oplus \mathbb{F}_p)$, where i is the canonical inclusion.

Definition 5.2.1 [EM]. A functor F is of degree k if and only if $\Delta^{k+1} F$ is trivial, but $\Delta^k F$ is non trivial.

Definition 5.2.2. An object F of $\mathcal{F}$ is said to be polynomial if it is of degree less or equal to k for some k. An object F is said to be analytic if it is the colimit of its polynomial subfunctors.

(5.2.3) Examples.

The tensor product $F \otimes G$ of a functor F of degree k with G a functor of degree ℓ is of degree $k + \ell$. This follows from the equivalence

$$\Delta(F \otimes G) \cong (\Delta F \otimes G) \oplus (F \otimes \Delta G) \oplus (\Delta F \otimes \Delta G).$$

The functor $V \mapsto V^{\otimes n}$ is polynomial of degree n.

The functor $V \mapsto H_n V$ is polynomial of degree n. Indeed, if $p = 2$, it is equivalent to the functor $V \mapsto (V^{\otimes n})^{\mathfrak{S}_n}$; if $p > 2$ it is equivalent to the functor

$$V \mapsto \bigoplus_{0 \leq 2i \leq n} \wedge^{n-2i}(V) \otimes (V^{\otimes i})^{\mathfrak{S}_i},$$

$V \mapsto \wedge^k V$ being the k-th exterior power functor.

Lemma 5.2.4. *The functor* $V \mapsto \mathbb{F}_p^{\mathrm{Hom}(V,W)}$ *is analytic for any* W.

This is part of Proposition 5.2.5 below, however there is a more direct and more elementary proof. First one observes that it is enough to prove the result for $W = \mathbb{F}_p$. Next, observe that the functor $V \mapsto V$ is a subfunctor of $V \mapsto \mathbb{F}_p^{V^*}$. This last functor takes values in the category of (p-Boolean) algebras. Thus one can consider the subfunctor $P_n(V)$ of $V \mapsto \mathbb{F}_p^{V^*}$ which is generated by n-fold products of elements of V and the unit. A dimension argument shows that

$$\mathrm{colim}_n P_n(V) = \mathbb{F}_p^{V^*};$$

in fact $P_n(V) = \mathbb{F}_p^{V^*}$ as soon as $n \geq (p - 1) \dim V$. Finally one checks that the functors P_n are polynomial of degree n. We refer to [**HLS2**] or to [**K1**] for more precisions.

It is possible to show that the functor $V \mapsto \mathbb{F}_p[V]$ is not analytic (Exercise).

Denote by $\mathcal{F}_n$ (resp. $\mathcal{F}_\omega$) the full subcategory of $\mathcal{F}$ of polynomial functors of degree less or equal to n (resp. of analytic functors). These are abelian subcategories of $\mathcal{F}$. They are stable under colimits.

Proposition 5.2.5. *For any unstable $\mathcal{A}$-module M the functor $f(M)$ is analytic. In particular the functor* $V \mapsto \mathbb{F}_p^{\mathrm{Hom}(V,W)}$ *is analytic for any* W.

Proof. For the first part because of Proposition 5.1.3, it is enough to check it on a set of generators of the category $\mathcal{U}$. But this is true for the unstable $\mathcal{A}$-module $F(n)$, for any n. For the second part observe that $V \mapsto \mathbb{F}_p^{\mathrm{Hom}(V,W)}$ is equivalent to $f(H^*W)$.

The functor $f : \mathcal{U}/\mathcal{N}il \to \mathcal{F}$ induces a functor from $\mathcal{U}/\mathcal{N}il$ to $\mathcal{F}_\omega$. Here is the main result of this chapter.

Theorem 5.2.6. *The functor induced by* f *from* $\mathcal{U}/\mathcal{N}il$ *to* $\mathcal{F}_\omega$ *is an equivalence of categories.*

The proof we shall give of this theorem is not the original one of [**HLS2**]. The proof here depends on internal properties of $\mathcal{F}_\omega$ rather than on the behaviour of f with respect to certain products. The proof implies that $\{H_n\}$, $n \in \mathbb{N}$, is a set of generators for $\mathcal{F}_\omega$. A direct proof of this fact has led Kuhn to a new proof of the $\mathcal{U}$-injectivity of H^*V, and to a new proof of the Adams-Gunawardena-Miller, Lannes theorem [**K1**].

We note also that the preceding result extends to the category $\mathcal{K}$. It leads to a new approach to the Adams-Wilkerson theorems [**AW**] and to the results of W. Rector and S. Lam as well as to generalizations. For these developments we refer to [**HLS1**] and [**HLS2**].

5.3. Injective objects in the category $\mathcal{F}_\omega$

In this section, we prove some technical facts, about $\mathcal{F}_\omega$, that we shall need for the proof of Theorem 5.2.6.

Lemma 5.3.1. *Let* W *be in* $\mathcal{E}_f$. *The functor* $E_W : V \mapsto \mathbb{F}_p^{\mathrm{Hom}(V,W)}$ *is injective in* $\mathcal{F}$. *It is a representing object for the functor* $F \mapsto F(W)^*$, $\mathcal{F} \to \mathcal{E}$.

Observe that if $W = \mathbb{F}_p$ the functor $E_{\mathbb{F}_p}$ decomposes canonically as the direct sum of the constant functor (included in $E_{\mathbb{F}_p}$ as the set of constant functions from V^* to $\mathbb{F}_p$) and of a functor $\widetilde{E}$. By the very definition $\mathrm{Hom}(F, \widetilde{E})$ is canonically isomorphic to $\Delta F(0)$. More generally $\mathrm{Hom}(F, \widetilde{E}^{\otimes k})$ is canonically isomorphic to $\Delta^k F(0)$. This will be useful later.

Corollary 5.3.2. *The set* $\{E_{\mathbb{F}_p^{\oplus n}}\}$, $n \in \mathbb{N}$, *is a set of cogenerators for the category* $\mathcal{F}$.

Recall that $\{C_n\}$, $n \in \mathbb{N}$, is a set of cogenerators for $\mathcal{F}$ if any object of $\mathcal{F}$ embeds in a product $\prod_\alpha C_{n_\alpha}$.

The corollary is a direct consequence of the lemma. Indeed, the product of the natural transformations $\gamma_x : F \mapsto E_W$, described below in the proof of Lemma 5.3.1, determines an embedding of F in the product

$$\prod_W \prod_{x \in F(W)^*} E_W .$$

Proof of Lemma 5.3.1. For any W in $\mathcal{E}_f$, let us describe a map $F(W)^* \mapsto \mathrm{Hom}_\mathcal{F}(F, E_W)$. To an element x in $F(W)^*$ we associate the natural transformation $\gamma_x : F \to E_W$ which sends $v \in F(V)$ to the map $\alpha \mapsto x(F(\alpha)(v))$, $\mathrm{Hom}(V, W) \to \mathbb{F}_p$. This determines a natural equivalence whose inverse is described as follows. To an element δ of $\mathrm{Hom}_\mathcal{F}(F, E_W)$ we associate the linear form $x \mapsto \delta_W(x)(\mathrm{Id}_W)$, $F(W) \to \mathbb{F}_p$, where $\delta_W : F(W) \to \mathbb{F}_p^{\mathrm{End}\ W}$ denotes the linear map induced by δ.

Thus the category $\mathcal{F}$ has enough injective objects and we can construct injective resolutions with products of the cogenerators E_W. The category $\mathcal{F}_\omega$ also has enough injectives. This will follow from Corollary 5.3.4.

Proposition 5.3.3. *The category $\mathcal{F}_\omega$ is locally noetherian.*

Recall that this means that the category has a set of noetherian generators.

Corollary 5.3.4. *Any direct sum $\oplus_\alpha E_{W_\alpha}$ is an injective object.*

Proof of Proposition 5.3.3. It is enough to show that any polynomial functor F such that $\dim_{\mathbb{F}_p} F(\mathbb{F}_p^{\oplus d})$ is finite for all integers d is noetherian. Indeed one observes that a 'set' of representatives of such functors is a set of generators for the category $\mathcal{F}_\omega$. This is proved as follows. Let $F \in \mathcal{F}$, $x \in F(V)$ for some V and denote by F_x the smallest sub-functor of F that contains x. Next $\dim_{\mathbb{F}_p} F_x(\mathbb{F}_p^{\oplus d})$ is finite for all d, indeed it is of dimension less than the cardinal of $\mathrm{Hom}(V, (\mathbb{F}_p^{\oplus d}))$. But F is the filtered colimit of the F_x's for all V all x. Restricting to analytic functors yields the result because an analytic functor is the filtered colimit of its polynomial subfunctors.

Define $P_F(d)$ to be the function $\dim_{\mathbb{F}_p} F(\mathbb{F}_p^{\oplus d})$. This function takes non-negative values on non-negative integers. We need:

Lemma 5.3.5. *Let F be in $\mathcal{F}_\omega$. Assume that $\dim_{\mathbb{F}_p} F(\mathbb{F}_p^{\oplus d})$ is finite for any d. The two following conditions are equivalent*

(i) $F \in \mathcal{F}_k$;

(ii) $P_F(d)$ *is a polynomial function of degree less than k.*

Given a function $P : \mathbb{N} \to \mathbb{Z}$ let ΔP be the function defined by the formula $\Delta P(d) = P(d+1) - P(d)$. Recall that a function $P : \mathbb{N} \to \mathbb{Z}$ is polynomial if and only if for k large enough $\Delta^k P$ is trivial. The proof of Lemma 5.3.5 is done by induction on k using the observation that $P_{\Delta F}(d) = \Delta P_F(d)$.

Let us return to the proof of Proposition 5.3.3. A functor F in $\mathcal{F}_\omega$ is said to be simple if it has no non-trivial subfunctors. It is supposed to be non-trivial itself. The following claim clearly implies Proposition 5.3.3.

A functor F in $\mathcal{F}_k$, such that $P_F(d)$ is finite for all d, has a finite filtration $0 = F_0 \subset F_1 \subset \dots \subset F_l = F$ whose quotient functors are simple.

One observes that a polynomial that takes non-negative integer values on non-negative integers cannot be the sum of an arbitrary large of many such non-trivial polynomials. This implies that F has only a finite number of non-trivial subquotients and yields the result. Alternatively it is possible to do a proof by induction on the degree of F observing that for a functor G, ΔG is trivial if and only if G is a constant functor.

Note that the above does not prove that all simple functors are polynomial. In fact this is true and follows from the fact that a simple functor is a subfunctor of some E_W.

We now show that $\mathcal{F}_\omega$ has enough injective objects. It is a consequence of the following Proposition-Definition. Recall that an object is semi-simple if it is isomorphic to a direct sum of simple objects.

Proposition-Definition 5.3.6. *Let F be an analytic functor. Then there exists an unique subfunctor $\mathrm{Soc}(F)$ of F (the socle of F) such that:*

(i) $\mathrm{Soc}(F)$ *is semi-simple;*

(ii)   *any simple subobject of F is contained in $\mathrm{Soc}(F)$;*

(iii) *any non-trivial subfunctor $G \subset F$ has a non-trivial 'intersection' with $\mathrm{Soc}(F)$.*

Proof.   The fact that $\mathrm{Soc}(F)$ exists, having properties (i) and (ii), is classical in abelian categories. It depends on Zorn's lemma. It remains to prove the last statement.

Let G be a subfunctor of F. As G is analytic it contains a non-trivial polynomial subfunctor F such that $\dim_{\mathbb{F}_p} F(\mathbb{F}_p^{\oplus d})$ is finite for all integers d. This functor contains a non-trivial simple subfunctor. Therefore G contains a simple subfunctor. Hence G and $\mathrm{Soc}(F)$ have a non-trivial intersection.

Corollary 5.3.7.   *An analytic functor F embeds in a direct sum $\oplus_\alpha E_{W_\alpha}$. Moreover if the socle of F is a finite direct sum of simple functors, F embeds in a finite direct sum $\oplus_\alpha E_{W_\alpha}$.*

Proof.   Write $\mathrm{Soc}(F)$ as $\oplus_\alpha S_\alpha$, where S_α is simple. The functor S_α embeds in E_{W_α} for W_α such that $S_\alpha(W_\alpha) \neq 0$. As $\oplus_\alpha E_{W_\alpha}$ is injective the embedding $\oplus_\alpha S_\alpha \to \oplus_\alpha E_{W_\alpha}$ extends to a morphism $F \to \oplus_\alpha E_{W_\alpha}$. It is an embedding as it is a monomorphism on $\mathrm{Soc}(F)$. The second part of the lemma follows immediately.

Before proving Theorem 5.2.6 we need some more facts on injective resolutions in $\mathcal{F}_\omega$.

Theorem 5.3.8.   *Let F be in $\mathcal{F}_k$, assume that $P_F(d)$ is finite for all d. There exists an injective resolution of F :*

$$0 \to F \to I_0 \to I_1 \to \dots \to I_k \to$$

such that all the functors I_k are finite direct sums of E_W's.

In fact we shall only need that I_0 and I_1 have this property. The proof depends on the following lemma.

Lemma 5.3.9.   *Let F be an analytic functor such that $\mathrm{Soc}(F)$ is a finite direct sum of simple functors and such that for some k $\Delta^k F$ is injective. Let $\gamma : F \to \oplus_\alpha E_{W_\alpha}$ be any embedding of F in a finite* *direct sum. Then the socle of the cokernel of* γ *is a finite direct sum of simple functors.*

Note that the condition holds for any polynomial subfunctor F. We conjecture that the condition, that $\Delta^k F$ is injective for some k, can be removed. (This condition does not hold for all functors.)

Proof of 5.3.9.   As Δ is exact we get an exact sequence

$$0 \to \Delta^k F \to \bigoplus_\alpha \Delta^k E_{W_\alpha} \to \Delta^k \operatorname{Coker} \gamma \to 0.$$

In this exact sequence the two first terms are injective. Thus the functor $\Delta^k \operatorname{Coker} \gamma$ is also injective and is a direct summand in $\oplus_\alpha \Delta^k E_{W_\alpha}$. Hence its socle is contained in the socle of $\oplus_\alpha \Delta^k E_{W_\alpha}$. But this last one is a finite direct sum. Indeed as $\operatorname{End} E_U \cong \mathbb{F}_p[\operatorname{End} U]$ is finite the socle of E_U is a finite direct sum. But one has $\Delta^k E_U \cong \mathbb{F}_p[U^{*\oplus k}] \otimes E_U$ and it follows that $\oplus_\alpha \Delta^k E_{W_\alpha}$ is a finite direct sum of E_U's. Hence the socle of $\Delta^k \operatorname{Coker} \gamma$ is a finite direct sum. It remains to deduce the result for $\operatorname{Coker} \gamma$ itself. This is shown as follows.

First, let S denote the socle of $\operatorname{Coker} \gamma$. The socle of $\Delta^k S$ is contained in the socle of $\Delta^k \operatorname{Coker} \gamma$. Next, there are only finitely many isomorphism classes of simple functors of degree less than a given integer k. Indeed by (5.3.1) and the comments which follow, these functors should appear non-trivially in the socle of E_W for $\dim W \leq k$, and the socle of this functor is a finite direct sum because it is injective and because its ring of endomorphisms is finite ((5.3.1) again). Moreover, ((5.3.1) again!), one shows that a simple functor S occurs as a subquotient of a given E_U with a finite multiplicity (this multiplicity being the dimension of $S(U)^*$). Hence the same is true for $\operatorname{Coker} \gamma$.

Thus the socle of $\operatorname{Coker} \gamma$ has only finitely many components in a given degree. As the operator $\Delta^k F$ is trivial only if F is of degree strictly less than k we are done.

The reader will find a somewhat improved version of this Theorem 5.3.8 in [**FLS**]. We have kept the original version here because of the following consequence. In the injective resolution of a polynomial functor of degree d it is possible to assume that all simple functors occurring in the socle of I_k are of degree less than $d(k+1)$, this estimate is accurate in low degrees for various cases. There is an analogous statement in the context of [**FLS**] but it is not accurate.

5.4. Proof of Theorem 5.2.6

In order to prove Theorem 5.2.6, we have to show that any object F of $\mathcal{F}_\omega$ is equivalent to an object of the form $f(M)$ for some M in $\mathcal{U}$. And we have to show also that any morphism $\gamma : F \to G$ is equivalent to one of the form $f(\varphi) : f(M) \to f(N)$. We are going to prove the first part. The second is proved in the same way.

By a colimit argument we can restrict attention to the case where F is polynomial of degree k and where $P_F(d)$ is finite for all d. Consider the two first terms of an injective resolution of F in $\mathcal{F}_\omega$:

$$0 \to F \to \bigoplus_\alpha E_{W_\alpha} \xrightarrow{\varphi} \bigoplus_\beta E_{U_\beta}$$

where each direct sum is finite. The natural transformation φ is an element of $\mathrm{Hom}_{\mathcal{F}}(\oplus_\alpha E_{W_\alpha}, \oplus_\beta E_{U_\beta}) \cong \underset{\alpha,\beta}{\oplus} \mathbb{F}_p[\mathrm{Hom}(U_\beta, W_\alpha)]$. It induces a map φ^* in cohomology:

$$\bigoplus_\alpha H^* W_\alpha \xrightarrow{\varphi^*} \bigoplus_\beta H^* U_\beta .$$

Let M be the unstable $\mathcal{A}$-module $\ker\varphi^*$. Apply the functor f to the exact sequence

$$0 \to M \to \bigoplus_\alpha H^* W_\alpha \xrightarrow{\varphi^*} \bigoplus_\beta H^* U_\beta$$

to get, using the Adams-Gunawardena-Miller, Lannes theorem, the exact sequence

$$0 \to f(M) \to \bigoplus_\alpha E_{W_\alpha} \to \bigoplus_\beta E_{U_\beta} .$$

Hence $f(M)$ is equivalent to F.

In fact, by uniqueness up to homotopy of the injective resolution, M is well defined up to isomorphism and the assignment $F \mapsto M$

determines a functor $\mathcal{F}_\omega \to \mathcal{U}/\mathcal{N}il$ which is right adjoint to f and which is 'the' inverse equivalence.

5.5. The functor $p_n : \mathcal{U}/\mathrm{Nil} \to \mathcal{M}\mathrm{od}_{\mathbb{F}_2[\mathfrak{S}_n]}$ and the filtration on $\mathcal{U}/\mathcal{N}il$

In this section we push one step further the analysis of the category $\mathcal{U}/\mathcal{N}il$. We define a filtration on $\mathcal{U}/\mathcal{N}il$, and we identify the quotients of the filtration with the categories of modular representations of the symmetric groups. We shall describe in the following section simple objects in $\mathcal{U}/\mathcal{N}il$. Then we shall get, as a corollary, Formula 4.3.1. We are going to work at the prime 2, and we will explain in a short section at the end of the chapter how to treat the case of an odd prime.

In order to state the main result of this section we need the notation $\mathcal{M}\mathrm{od}_{\mathbb{F}_2[\mathfrak{S}_n]}$ for the category of right $\mathbb{F}_2[\mathfrak{S}_n]$-modules.

Recall that $\mathrm{Hom}_{\mathcal{U}}(M, K(1)^{\otimes n})$ is, in a natural way, a profinite $\mathbb{F}_2$-vector space and that the action by 'permutation of the factors' of $\mathfrak{S}_n$, $n \geq 0$ on $K(1)^{\otimes n}$ induces on $\mathrm{Hom}_{\mathcal{U}}(M, K(1)^{\otimes n})$ a (left) $\mathfrak{S}_n$-action. (Note that $K(1)^{\otimes 0} \cong \mathbb{F}_2$.) Thus the continuous dual $\mathrm{Hom}_{\mathcal{U}}(M, K(1)^{\otimes n})'$ is a right $\mathbb{F}_2[\mathfrak{S}_n]$-module. Hence the assignment

$$M \mapsto \mathrm{Hom}_{\mathcal{U}}(M, K(1)^{\otimes n})'$$

determines a functor $p_n : \mathcal{U} \to \mathcal{M}\mathrm{od}_{\mathbb{F}_2[\mathfrak{S}_n]}$. By the very construction the functor p_n commutes with colimits. It is right exact. As $K(1)^{\otimes n}$ is injective it is left exact. Hence it is exact. It is trivial on the subcategory $\mathcal{N}il$ and it factors uniquely as:

$$\mathcal{U} \to \mathcal{U}/\mathcal{N}il \to \mathcal{M}\mathrm{od}_{\mathbb{F}_2[\mathfrak{S}_n]} .$$

This defines (by abuse of notation) the functor

$$p_n : \mathcal{U}/\mathcal{N}il \to \mathcal{M}\mathrm{od}_{\mathbb{F}_2[\mathfrak{S}_n]} .$$

Next, define the category $\mathcal{V}_n$ to be the full sucategory of $\mathcal{U}/\mathcal{N}il$ consisting of objects on which the functor induced by p_k is trivial for all $k > n$. By definition the functor p_n is trivial on $\mathcal{V}_{n-1}$

Theorem 5.5.1 (Compare with [LS4] and [FS]). *The functor p_n induces an equivalence of categories* $\mathcal{V}_n/\mathcal{V}_{n-1} \to \mathcal{M}\mathrm{od}_{\mathbb{F}_2[\mathfrak{S}_n]}$.

The functor $p_n : \mathcal{V}_n/\mathcal{V}_{n-1} \to \mathrm{Mod}_{\mathbb{F}_2[\mathfrak{S}_n]}$ is faithful by definition. To complete the proof of the theorem we have to prove it is full and that for any object R of $\mathrm{Mod}_{\mathbb{F}_2[\mathfrak{S}_n]}$ there exists an object M in $\mathcal{V}_n$ such that $p_n(M) \cong R$. We are going to prove the last assertion, the other one is proved similarly.

We shall need the following lemma due to Carlsson [Cl1]. Consider the map $\mu_{a_1,\ldots,a_n} : K(1)^{\otimes n} \to K(2^{a_1} + \ldots + 2^{a_n})$ which is the following composite:

$$K(1)^{\otimes n} \xrightarrow{\cong} \bigotimes_{i=1,\ldots,n} K(2^{a_i}) \xrightarrow{\mathrm{mult}} K(2^{a_1} + \ldots + 2^{a_n}).$$

Lemma 5.5.2. *The connectivity of the map* $\mu_{a_1,\ldots,a_n}$ *tends to infinity with* $\inf_{i \neq j} |a_i - a_j|$.

Proof. This is an exercise using the structure theorem for K_*^* (see Chapter 4).

We shall need also.

Lemma 5.5.3. *The representation* $p_n\big(F(1)^{\otimes n}\big)$ *is isomorphic to* $\mathbb{F}_2[\mathfrak{S}_n]$.

Lemma 5.5.4. *The ring* $\mathrm{End}_{\mathcal{U}}\big(F(1)^{\otimes n}\big)$ *is isomorphic to* $\mathbb{F}_2[\mathfrak{S}_n]$.

The first lemma is a direct consequence of the definition of p_n and of Lemma 5.5.2.

The second follows from:

— a map $\varphi : F(1)^{\otimes n} \to F(1)^{\otimes n}$ is known as soon as it is known on $u \otimes u^2 \otimes \ldots \otimes u^{2^{n-1}}$;

— the element $\varphi(u \otimes u^2 \otimes \ldots \otimes u^{2^{n-1}})$ can be written uniquely as $s(u \otimes u^2 \otimes \ldots \otimes u^{2^{n-1}})$, for some $s \in \mathbb{F}_2[\mathfrak{S}_n]$.

The first fact follows from the observation that the sub-module of $F(1)^{\otimes n}$ generated by $u \otimes \ldots \otimes u^{2^{n-1}}$ is isomorphic in $\mathcal{U}/\mathcal{N}il$ to $F(1)^{\otimes n}$. Indeed for any class $z \in F(1)^{\otimes n}$ there exists an integer $k \in \mathbb{N}$ and an operation $\theta \in \mathcal{A}$ such that $\theta(u \otimes \ldots \otimes u^{2^{n-1}}) = \mathrm{Sq}_0^k z$.

Let us come back to the proof of (5.5.1). By a colimit argument we can restrict attention to the case of a finitely generated $\mathbb{F}_2[\mathfrak{S}_n]$-

module R. Consider the beginning of a projective resolution of such a module:

$$\mathbb{F}_2[\mathfrak{S}_n]^{\oplus\beta} \xrightarrow{\varphi} \mathbb{F}_2[\mathfrak{S}_n]^{\oplus\alpha} \to R \to 0.$$

The direct sums are finite. Using Lemma 5.5.3 and Lemma 5.5.4 observe that

$$\mathrm{Hom}_{\mathbb{F}_2[\mathfrak{S}_n]}\left(\mathbb{F}_2[\mathfrak{S}_n]^{\oplus\beta}, \mathbb{F}_2[\mathfrak{S}_n]^{\oplus\alpha}\right) \cong \left(\mathbb{F}_2[\mathfrak{S}_n]^{\oplus\alpha}\right)^{\times\beta}$$

$$\cong \mathrm{Hom}_{\mathcal{U}}\left(\left(F(1)^{\otimes n}\right)^{\oplus\beta}, \left(F(1)^{\otimes n}\right)^{\oplus\alpha}\right).$$

Denote by $\bar{\varphi}$ the element of $\mathrm{Hom}_{\mathcal{U}}((F(1)^{\otimes n})^{\oplus\beta}, (F(1)^{\otimes n})^{\oplus\alpha})$ corresponding to φ under this isomorphism.

The unstable module $\mathrm{Coker}\,\bar{\varphi}$ is easily checked to be in $\mathcal{V}_n$. A routine check shows that $p_n(\mathrm{Coker}\,\bar{\varphi})$ is isomorphic to R. It follows that the functor induced by p_n:

$$\mathcal{V}_n/\mathcal{V}_{n-1} \to \mathcal{M}\mathrm{od}_{\mathbb{F}_2[\mathfrak{S}_n]}$$

is an equivalence of categories. In fact, by uniqueness up to homotopy of the injective resolution, $\mathrm{Coker}\,\bar{\varphi}$ is well defined up to isomorphism and the assignment $R \mapsto \mathrm{Coker}\,\bar{\varphi}$ determines, by composition with the forgetful functor, a functor $\mathcal{M}\mathrm{od}_{\mathbb{F}_2[\mathfrak{S}_n]} \to \mathcal{V}_n/\mathcal{V}_{n-1}$ which is left adjoint to the functor induced by p_n and which is 'the' inverse equivalence. We shall denote it by m_n.

We are going to give now a 'concrete' description of the category $\mathcal{V}_n$. This will make the link with [LS4] and [FS].

Proposition 5.5.5. *The category $\mathcal{V}_n$ is the full subcategory of $\mathcal{U}/\mathcal{N}il$ consisting of objects M such that:*

$$M \in \mathcal{V}_n \Leftrightarrow s_1(M) \text{ is trivial in degrees } d \text{ such that } \alpha(d) > n.$$

Recall that $\alpha(d)$ is the number of 1's in the 2-adic expansion of d and that $s_1 : \mathcal{U}/\mathcal{N}il \to \mathcal{U}$ is the right adjoint of the forgetful functor $r = r_1 : \mathcal{U} \to \mathcal{U}/\mathcal{N}il$. Observe that it would be equivalent to require that any *reduced* unstable $\mathcal{A}$-module L which is isomorphic to M in $\mathcal{U}/\mathcal{N}il$ has the property stated for $s_1 M$.

Call $\mathcal{V}'_n$ the full subcategory of $\mathcal{U}/\mathcal{N}il$ defined by the condition of Proposition 5.5.5. We are going to identify it with $\mathcal{V}_n$. In view of the comment above it is easy to show, using (5.5.2), that the category

$\mathcal{V}_n$ contains the category $\mathcal{V}'_n$. Let $\mathcal{V}''_{n-1}$, $n > 0$ denote the full subcategory of $\mathcal{V}_n$ which is the 'kernel' of the functor p_n restricted to $\mathcal{V}'_n$. It contains $\mathcal{V}_{n-1}$ and is the intersection of the categories $\mathcal{V}_n$ and $\mathcal{V}'_{n-1}$. One has $\mathcal{V}_n \supset \mathcal{V}''_n \supset \mathcal{V}'_n$.

One observes that exactly the same proof as above yields an equivalence of categories:

$$\mathcal{V}'_n/\mathcal{V}''_{n-1} \to \mathcal{M}od_{\mathbb{F}_2[\mathfrak{S}_n]} \ .$$

The inverse equivalence, denoted also by m_n is described as above. Next, we are going to identify the categories $\mathcal{V}_{n-1}$, $\mathcal{V}'_{n-1}$ and $\mathcal{V}''_{n-1}$. This will complete the proof of (5.5.5). We need:

Lemma 5.5.6. *Let M be an unstable module such that $p_n(M)$ is trivial for all $n \geq 0$. Then M is nilpotent.*

It follows from Chapter 3, that any $L \in \mathcal{L}$ is a direct summand of $K(1)^{\otimes n}$ for some n. Indeed L is a direct summand of $\bar{H}^{\otimes n}$ for some n (see 3.12) which is itself a direct summand of $K(1)^{\otimes n}$ (see 3.1). Thus 3.14.3 implies the lemma. The lemma says that '$\mathcal{V}_{-1}$ is empty'.

It is clear in view of the lemma and the above equivalences that the categories $\mathcal{V}''_0$, $\mathcal{V}'_0$ and $\mathcal{V}_0$ are the same. Thus we assume inductively that $\mathcal{V}''_{n-1}$, $\mathcal{V}'_{n-1}$ and $\mathcal{V}_{n-1}$ are the same and prove it for n. Consider an object M in $\mathcal{V}_n$ and the unit of the adjunction $\eta : m_n p_n M \to M$. By the very construction the unstable module $m_n p_n M$ is trivial in $\mathcal{V}'_n$, and $p_n(\eta)$ is an isomorphism. It follows that the kernel and cokernel of η are in $\mathcal{V}_{n-1}$ and thus in $\mathcal{V}'_{n-1}$. We are done.

We note that, by definition, a reduced representative in $\mathcal{U}$ of an object in $\mathcal{V}_n$ is a filtered colimit of reduced objects in $\mathcal{U}$ such that M is trivial in degrees d that satisfy the two following conditions:

(i) $\alpha(d) > n$; and

(ii) if $2^{a_1} + ... + 2^{a_t}$ is the 2-adic expansion of d the integer $\inf_{i \neq j}|a_i - a_j|$ is larger than a fixed integer depending only on M.

The preceding conditions means that the $\mathbb{F}_2$-vector space, $p_k(M)$ isomorphic to $M^{2^{a_1} + ... + 2^{a_k}}$ (M finitely generated) is trivial as soon as $k > n$ and $\inf_{i \neq j}|a_i - a_j|$ is large enough. Our result shows that, in fact, these vector spaces are trivial as soon as $k > n$ i.e when condition (i) holds.

Remark. Theorem 5.5.1 was originally proved by Lannes and the author [**LS4**] using the main result of Chapter 5 and a 'characteristic p version' of the classical analysis of analytic functors [**Md**]. Note also that the functor p_n introduced here agrees on $\mathcal{V}_n$ with the one introduced in [**FS**] but that these two functors do not agree on the whole of $\mathcal{U}/\mathcal{N}il$.

To conclude this section we compare the categories $\mathcal{V}_n$ and $\mathcal{F}_n$.

Proposition 5.5.7. *The equivalence* $f : \mathcal{U}/\mathcal{N}\mathrm{il} \to \mathcal{F}_\omega$ *restricts to an equivalence* $\mathcal{V}_n \to \mathcal{F}_n$

We just outline of the proof, leaving the details to the reader. The functor $\bar{T}$ induces a functor on the category $\mathcal{U}/\mathcal{N}il$. Indeed, in order to apply the universal property of $\mathcal{U}/\mathcal{N}il$, it is enough to check that if M is nilpotent $\bar{T}(M)$ is again nilpotent. But this is true by direct computation for the $H^*V \otimes J(n)$'s, $n > 0$ and the general case follows from 3.11.

We claim that f induces a functor from $\mathcal{V}_n$ to $\mathcal{F}_n$. This is proved by induction on n by using the following facts

— for all M in $\mathcal{U}/\mathcal{N}il$ one has

$$f(\bar{T}(M) \cong \Delta f(M) ;$$

— a noetherian object of $\mathcal{V}_n$ is isomorphic in the quotient category $\mathcal{V}_n/\mathcal{V}_{n-1}$ to the cokernel of a certain map

$$\bar{\varphi} \in \mathrm{Hom}_\mathcal{U}\left(\left(F(1)^{\otimes n}\right)^{\oplus \beta}, \left(F(1)^{\otimes n}\right)^{\oplus \alpha}\right)$$

(see the proof of 5.5.1);

— the known action of $\bar{T}$ on $F(1)^{\otimes n}$.

The first two facts imply that $\bar{T}(\mathcal{V}_n) \subset \mathcal{V}_{n-1}$.

The functor from $\mathcal{V}_n$ to $\mathcal{F}_n$ is full and faithful because f is full and faithful. It remains to prove that any object in $\mathcal{F}_n$ is isomorphic to an object of the form $f(M)$, $M \in \mathcal{V}_n$. It is enough to check that if M belongs to $\mathcal{V}_n - \mathcal{V}_{n-1}$, then $\bar{T}^n(M)$ is non-trivial. Recall that a noetherian object M of $\mathcal{V}_n$ is isomorphic in the quotient category $\mathcal{V}_n/\mathcal{V}_{n-1}$ to the cokernel of a certain map

$$\bar{\varphi} \in \mathrm{Hom}_\mathcal{U}\left(\left(F(1)^{\otimes n}\right)^{\oplus \beta}, \left(F(1)^{\otimes n}\right)^{\oplus \alpha}\right) .$$

Using the fact that $\mathbb{F}_2[\mathfrak{S}_n]$ is injective in the category of $\mathbb{F}_2[\mathfrak{S}_n]$-modules one shows that it is also isomorphic to the kernel of a map

$$\bar{\psi} \in \text{Hom}_{\mathcal{U}} \left(\left(F(1)^{\otimes n} \right)^{\oplus \beta}, \left(F(1)^{\otimes n} \right)^{\oplus \alpha} \right) .$$

Hence if we suppose that M is non-trivial and in $\mathcal{V}_n - \mathcal{V}_{n-1}$ it is obviously equipped with a non-trivial map to $(\tilde{H}^*\mathbb{Z}/2)^{\otimes n}$ induced by the embedding $F(1)^{\otimes n} \to (\tilde{H}^*\mathbb{Z}/2)^{\otimes n}$. The result follows.

5.6. Simple objects of $\mathcal{U}/\mathcal{N}il$

In order to prove Formula 4.3.1 we need a more precise description of simple objects of $\mathcal{U}/\mathcal{N}il$ in terms of representation theory of the symmetric groups. Recall that by a simple object we mean a non-trivial object that has no non-trivial subobjects. The next two propositions together with the remark following the first one give the complete description of simple objects of the category $\mathcal{U}/\mathcal{N}il$.

Let S be a simple object of $\mathcal{U}/\mathcal{N}il$ that is in $\mathcal{V}_n$ but not in $\mathcal{V}_{n-1}$. The following proposition is a direct consequence of the results in Section 5.5.

Proposition 5.6.1. *The representation* $p_n(S)$ *is simple.*

As the unstable $\mathcal{A}$-modules $F(k)$ are generators for the category $\mathcal{U}/\mathcal{N}il$ and as $F(k)$ is in $\mathcal{V}_k$ any simple object S of $\mathcal{U}/\mathcal{N}il$ has to be in $\mathcal{V}_n$ for some n. The smallest abelian subcategory of $\mathcal{U}/\mathcal{N}il$ which contains all $\mathcal{V}_k$'s and is stable under colimit is $\mathcal{U}/\mathcal{N}il$ itself.

Let R be a representation of the symmetric group $\mathfrak{S}_n$. Recall the norm map

$$R \otimes F(1)^{\otimes n} \xrightarrow{N} R \otimes F(1)^{\otimes n}$$

which is given by

$$N(r \otimes u^{2^{a_1}} \otimes \dots \otimes u^{2^{a_n}}) = \sum_{\sigma \in \mathfrak{S}_n} \sigma(r) \otimes u^{2^{a_{\sigma(1)}}} \otimes \dots \otimes u^{2^{a_{\sigma(n)}}} .$$

Proposition 5.6.2. *Let* R *be a simple* $\mathbb{F}_2[\mathfrak{S}_n]$*-module. Then the image of the norm map* N

$$R \otimes F(1)^{\otimes n} \xrightarrow{N} R \otimes F(1)^{\otimes n}$$

is a simple object of $\mathcal{V}_n - \mathcal{V}_{n-1}$. *Any simple object of* $\mathcal{V}_n - \mathcal{V}_{n-1}$ *is isomorphic to one of this form. The representation* $p_n(ImN)$ *is equivalent to the representation* R.

Moreover let $e \in \mathbb{F}_2[\mathfrak{S}_n]$ *be such that* $e\,\mathbb{F}_2[\mathfrak{S}_n]$ *is isomorphic to* R. *Then the image of the norm map is isomorphic, as an unstable* $\mathcal{A}$*-module, to* $eF(1)^{\otimes n}$.

Remark. Propositions 5.6.1 and 5.6.2 are an easy consequence of the approach of [**LS4**]. It is also possible to prove them directly, this was done by the author using an approach closely related to [**FS**].

Proof of Proposition 5.6.2. We shall need the following

Lemma 5.6.3 [FS]. *The largest subobject of* $F(1)^{\otimes n}$ *which is in* $\mathcal{V}_{n-1}$ *is trivial.*

The proof is very similar to some of earlier chapters and will be given later.

Let $R \cong e\,\mathbb{F}_2[\mathfrak{S}_n]$ be simple. First let us show that $eF(1)^{\otimes n}$ is simple in $\mathcal{U}/\mathcal{N}il$. The unstable $\mathcal{A}$-module $eF(1)^{\otimes n}$ does not contain a non-trivial sub-object of $\mathcal{V}_{n-1}$ by Lemma 5.6.3. It follows from Proposition-Definition 5.3.6 that there exists a simple subobject, S, of $eF(1)^{\otimes n}$ which is in $\mathcal{V}_n - \mathcal{V}_{n-1}$. This could be proved directly by using the techniques of the preceding section (exercise, see also [**FS**]). Consider the quotient $eF(1)^{\otimes n}/S$, it is an object of $\mathcal{V}_{n-1}$. Indeed, $p_n(eF(1)^{\otimes n}) \cong e\,\mathbb{F}_2[\mathfrak{S}_n] \cong R$ is simple as well as $p_n(S)$. Thus $R \cong p_n(S)$ and the representation $p_n(eF(1)^{\otimes n}/S)$ is trivial.

Next, let us show that $eF(1)^{\otimes n}/S$ is trivial. A priori it belongs to $\mathcal{V}_{n-1}$. Observe now that the class $e(u \otimes u^2 \otimes ... \otimes u^{2^{n-1}})$ is a generator of $eF(1)^{\otimes n}$ as an object of $\mathcal{U}/\mathcal{N}il$. Formally this means that the sub-module of $eF(1)^{\otimes n}$ generated by the class $e(u \otimes ... \otimes u^{2^{n-1}})$ is isomorphic, in $\mathcal{U}/\mathcal{N}il$, to $eF(1)^{\otimes n}$. Indeed, the sub-module of $F(1)^{\otimes n}$ generated by $u \otimes ... \otimes u^{2^{n-1}}$ is isomorphic in $\mathcal{U}/\mathcal{N}il$ to $F(1)^{\otimes n}$ because for any class $z \in F(1)^{\otimes n}$ there exists an integer

$n \in \mathbb{N}$ and an operation $\theta \in \mathcal{A}$ such that $\theta(u \otimes \dots \otimes u^{2^{n-1}}) = \mathrm{Sq}_0^n z$. Thus the same holds for the class $e(u \otimes \dots \otimes u^{2^{n-1}})$ in $eF(1)^{\otimes n}$. This implies that $eF(1)^{\otimes n}/S$ is trivial. In fact, as the integer $\alpha(|e(u \otimes \dots \otimes u^{2^{n-1}})|)$ is equal to n and because of (5.5.5) the class $e(u \otimes \dots \otimes u^{2^{n-1}})$ has a trivial image in $eF(1)^{\otimes n}/S$, and consequently belongs to S. Hence $eF(1)^{\otimes n}$ is simple.

We now identify the image of the norm with $eF(1)^{\otimes n}$. This provides a link with [FS]. Let $\lambda : R \to \mathbb{F}_2$ be a non-zero linear form. For a non-zero element $r_0 \in R$, let $N(r_0)$ in $R \otimes \mathbb{F}_2[\mathfrak{S}_n]$ be the element $\sum_{\sigma \in \mathfrak{S}_n} r_0 \sigma \otimes \sigma^{-1}$. Finally let $\varepsilon \in \mathbb{F}_2[\mathfrak{S}_n]$ be $\sum_\sigma \lambda(r_0 \sigma)\sigma^{-1}$. We claim that R is isomorphic to $\varepsilon \mathbb{F}_2[\mathfrak{S}_n]$. Indeed consider the composite

$$R \to R \otimes \mathbb{F}_2[\mathfrak{S}_n] \xrightarrow{N} R \otimes \mathbb{F}_2[\mathfrak{S}_n] \xrightarrow{\lambda \otimes \mathrm{Id}} \mathbb{F}_2[\mathfrak{S}_n],$$

where the first map sends r to $r \otimes \mathrm{Id}$ and the norm map N sends $r \otimes \tau$ to $\Sigma r\sigma \otimes \sigma^{-1}\tau$. One checks that it yields an isomorphism between R and $\varepsilon \mathbb{F}_2[\mathfrak{S}_n]$. Replacing $\mathbb{F}_2[\mathfrak{S}_n]$ by $F(1)^{\otimes n}$ we get a map φ from the image of the norm to $\varepsilon F(1)^{\otimes n}$ that is easily shown to be surjective because $\varepsilon F(1)^{\otimes n}$ is simple in $\mathcal{U}/\mathcal{N}il$. It is injective because $p_n(\varphi)$ is an isomorphism and because of (5.6.3) and (5.5.5).

Proof of Lemma 5.6.3. It is enough to show that if a submodule M of $F(1)^{\otimes n}$ it is non-trivial in a degree d with $\alpha(d) < n$ it is also non-trivial in a degree d' with $\alpha(d') > \alpha(d)$. For the moment we shall do the following assumption on M:

$$z \in F(1)^{\otimes n} \text{ and } \mathrm{Sq}_0 z \in M \Rightarrow z \in M.$$

This condition will be explained in Chapter 6. Let x be in M^d, $x \neq 0$, $\alpha(d) < n$. Assume that x does not write as $\mathrm{Sq}_0 z$ for some $z \in M$. Consider all monomials $u^{2^{a_1}} \otimes \dots \otimes u^{2^{a_n}}$ adding up to x:

$$x = \sum_{(a_1, \dots, a_n)} u^{2^{a_1}} \otimes \dots \otimes u^{2^{a_n}}.$$

As $\alpha(d) < n$, in all of these monomials, two powers of u (at least) have to be equal. Suppose that u occurs at least twice in one of these monomials. Then the element $Q_i Q_j x$ (recall that the operations Q_i are Milnor's derivations), $i \neq j$, i and j large enough, is non-zero and such that $\alpha(|Q_i Q_j x|) > \alpha(|x|)$. Suppose now that x is not of the form $\mathrm{Sq}_0 z$ and that u occurs only one time in all monomials adding up to x. Apply the operation Q_i, $i \geq 0$. One has $Q_i x \neq 0$ and $\alpha(|Q_i(x)|) = \alpha(|x|)$. There exists x_1 in M such that $Q_i x = \mathrm{Sq}_0^k x_1$ for some k and x_1 is not of the form $\mathrm{Sq}_0 z$.

As x_1 is not of the form $\mathrm{Sq}_0 z$ in the monomials adding up to x_1 some factor as to be u. One applies the preceding process to x_1. If there is a monomial, among those adding up to x_1, in which u appears at least twice, we are done. Otherwise we get a class x_2 and we continue the process. Recall that since $\alpha(d) > n$ some power u^{2^a} has to occur at least two times in one of the monomial adding up to x. Hence we are done after a finite number of steps.

We now explain how to proceed for a general M. Let M' be the submodule of $F(1)^{\otimes n}$ of elements z such that there exits k for which $\mathrm{Sq}_0^k z \in M$. It satisfies the assumption we made at the beginning of the proof. Therefore there exists a non-zero element $z \in M'$ such that $\alpha(z) = n$. Now for k large enough $\mathrm{Sq}_0^k z \in M$.

5.7. Proof of Formula 4.3.1

Let R be a simple $\mathbb{F}_2[\mathfrak{S}_n]$-module, S be the associated simple object of $\mathcal{V}_n$, E_S the injective hull of S considered as an object of $\mathcal{U}$. Then

Theorem 5.7.1. *The number of occurrences of E_S in the decomposition of $K(i)$ is given by*

$$\sum_\mu s(i, \mu) \ \dim_{\mathbb{F}_2} \bar{\mathfrak{S}}_\mu R \,,$$

where $\mu = (\mu_1, ..., \mu_t)$ runs through the set of all partitions of n, $s(i, \mu)$ is the number of ways to write the integer i as $\Sigma \mu_\ell 2^{q_l}$, with $q_\ell \in \mathbb{Z}$ and $\ell \neq \ell' \Rightarrow q_\ell \neq q_{\ell'}$; $\mathfrak{S}_\mu \subset \mathfrak{S}_n$ is a Young subgroup isomorphic to $\mathfrak{S}_{\mu_1} \times ... \times \mathfrak{S}_{\mu_t}$, and $\bar{\mathfrak{S}}_\mu = \Sigma_\sigma \sigma \in \mathbb{F}_2[\mathfrak{S}_n]$.

There is a unique (up to isomorphism) direct sum decomposition of the unstable $\mathcal{A}$-module $K(i)$ of the form $E_S^{\oplus a_S} \oplus I$, such that E_S is not a direct summand of I. The integer a_S is well defined. We have

(5.7.2) $a_S = \dim_{\mathbb{F}_2} \mathrm{Hom}_\mathcal{U}(S, K(i)) \,.$

The Formula 5.7.2 is implied by the following two formulas

(5.7.3) $\dim_{\mathbb{F}_2} \mathrm{Hom}_{\mathcal{U}}(S, K(i)) = a_S \dim_{\mathbb{F}_2} \mathrm{End}_{\mathcal{U}}(S)$;

(5.7.4) $\dim_{\mathbb{F}_2} \mathrm{End}_{\mathcal{U}}(S) = 1$.

Formula 5.7.3 follows from the fact that $\dim_{\mathbb{F}_2} \mathrm{Hom}_{\mathcal{U}}(S, I)$ is zero which follows from the fact that E_S is not a direct summand of I. Then $\dim_{\mathbb{F}_2} \mathrm{Hom}_{\mathcal{U}}(S, E_S) \cong \dim_{\mathbb{F}_2} \mathrm{End}_{\mathcal{U}}(S)$ as S is the socle, in the category $\mathcal{U}/\mathcal{N}il$, of E_S.

Formula 5.7.4 depends on group representation theory. Observe, using 5.5, that

$$\mathrm{End}_{\mathcal{U}}(S) \cong \mathrm{End}_{\mathbb{F}_2[\mathfrak{S}_n]}(R)$$

and the result is classical [JK], indeed the simple $\mathbb{F}_2[\mathfrak{S}_n]$-modules are absolutely irreducible [CR][S2]. This also follows from Theorem 4.4.2 that says that S is a cyclic $\mathcal{A}$-module. Either one uses the formulas given for ε (in [J] or [JK]) in terms of the 2-regular partition $\lambda = (\lambda_1, ..., \lambda_d)$ determined by S. One observes that S is of dimension 1 in degree $\lambda_1 + 2\lambda_2 + ... + 2^{d-1}\lambda_d$ (and trivial in degrees less than that). This implies the result because a morphism $\varphi : S \to S$ is completely determined by its effect in this degree.

Next, by definition of $K(i)$ we get $a_S = \dim S^{2^q i}$, for q large enough. Then observe that, for q large enough, $\left(F(1)^{\otimes n}\right)^{2^q i}$ is isomorphic to the direct sum

$$\bigoplus_{\mu} \mathbb{F}_2 \left[\mathfrak{S}_n/\mathfrak{S}_\mu\right]^{\oplus s(i,\mu)} .$$

Indeed the monomial

$$\underbrace{u^{2^{q_1}} \otimes ... \otimes u^{2^{q_1}}}_{\mu_1} \otimes ... \otimes \underbrace{u^{2^{q_l}} \otimes ... \otimes u^{2^{q_l}}}_{\mu_l}$$

$(k \neq k' \Rightarrow q_k \neq q_{k'})$ generates an $\mathbb{F}_2[\mathfrak{S}_n]$-module isomorphic to the permutation module $\mathbb{F}_2[\mathfrak{S}_n/\mathfrak{S}_\mu]$.

Finally, it remains to compute the dimension of the image of the following norm map:

$$N : R \otimes \mathbb{F}_2[\mathfrak{S}_n/\mathfrak{S}_\mu] \to R \otimes \mathbb{F}_2[\mathfrak{S}_n/\mathfrak{S}_\mu] .$$

We leave it to the reader to check it is $\dim_{\mathbb{F}_2} \bar{\mathfrak{S}}_\mu R$.

5.8. Comments on the Weyl 'correspondence'

In Chapter 4 we described how to associate to an element in $\mathcal{L}_d$ a certain simple representation of $\mathrm{GL}_d\,\mathbb{F}_2$. Using the preceding results one can associate to any element of $\mathcal{L}$ a simple representation of a certain symmetric group. It is done as follows. Let L be in $\mathcal{L}$, as an object of $\mathcal{U}/\mathcal{N}il$ it contains a unique (non-trivial) simple subobject (see Section 5.6). This simple subobject S is determined by a simple representation of a certain symmetric group. In turns out that L is determined by S as it is its injective hull. Therefore we get a correspondence between simple representations of the general linear groups and simple representations of the symmetric groups. In fact this is nothing else than the classical correspondence of Weyl modules to Specht modules. Let us prove it.

Let $\lambda = (\lambda_1, ..., \lambda_k)$ be a partition of n, $\lambda' = (\lambda'_1, ..., \lambda'_t)$ be the conjugate partition ($\lambda_1 = t$). Assume that $\lambda_i > \lambda_{i+1}$, $i = 1, ..., d-1$. Consider the unstable module $\wedge^{\lambda_1} F(1) \otimes ... \otimes \wedge^{\lambda_d} F(1)$.

On one hand $f\big(\wedge^{\lambda_1} F(1) \otimes ... \otimes \wedge^{\lambda_d} F(1)\big)(V)$ is equivalent to $\wedge^{\lambda_1} V \otimes ... \otimes \wedge^{\lambda_d} V$ which is non-zero as soon as $\dim V \geq \lambda_1$. The Weyl module of $\mathrm{GL}_{\lambda_1}\,\mathbb{F}_2$ associated to λ' is a submodule (for $V = \mathbb{F}_2^{\oplus \lambda_1}$) therein. The simple representation of $\mathrm{GL}_{\lambda_1}\,\mathbb{F}_2$ associated to λ' is a certain subquotient of $\wedge^{\lambda_1}\big(\mathbb{F}_2^{\oplus \lambda_1}\big) \otimes ... \otimes \wedge^{\lambda_d}\big(\mathbb{F}_2^{\oplus \lambda_1}\big)$.

On the other hand $\mathrm{Hom}_{\mathcal{U}}(\wedge^{\lambda_1} F(1) \otimes ... \otimes \wedge^{\lambda_d} F(1), K(1)^{\otimes n})$ is isomorphic as an $\mathbb{F}_2[\mathfrak{S}_n]$-module to $\mathbb{F}_2[\mathfrak{S}_n/\mathfrak{S}_\lambda]$ and the Specht module associated to λ is a submodule of $\mathbb{F}_2[\mathfrak{S}_n/\mathfrak{S}_\lambda]$. The simple representation associated to λ is a subquotient therein.

Let us be more precise. Let $\lambda = (\lambda_1, ..., \lambda_k)$ be a 2-regular partition of n. It follows from [**JK**] that the Specht module associated is isomorphic to $\bar{R}_\lambda \bar{C}_\lambda \mathbb{F}_2[\mathfrak{S}_n]$ (here $\bar{R}_\lambda$, resp. $\bar{C}_\lambda$, denotes the sum of the elements of the rows stabilizer, resp. of the columns stabilizer, of the associated Young diagram). As $\lambda = (\lambda_1, ..., \lambda_k)$ is 2-regular it has a unique maximal non-trivial, hence simple, submodule. Let us call it D_λ, it is isomorphic to $\bar{C}_\lambda \bar{R}_\lambda \bar{C}_\lambda \mathbb{F}_2[\mathfrak{S}_n]$ (see [**JK**] or [**J**]). We note that the statements of the last reference are stated in terms of column 2-regular partitions, however the translation is easy.

The Weyl module $W_{\lambda'}$ associated to the column 2-regular partition λ' is isomorphic to $\bar{C}_{\lambda'} \bar{R}_{\lambda'} V^{\otimes n}$, where $V = \mathbb{F}_2^{\oplus \lambda_1'}$, see the Theorem 3.2 in [J]. The simple representation of $\mathrm{GL}_{\lambda_1'} \, \mathbb{F}_2$ attached to λ' is isomorphic to $\bar{R}_{\lambda'} \bar{C}_{\lambda'} \bar{R}_{\lambda'} V^{\otimes n}$. It is the minimal, hence simple, non-trivial subquotient of $W_{\lambda'}$ considered has a module over the hyperalgebra $\mathcal{U}_{\mathbb{F}_2}$ (see [JK] or [J]). As λ' is column 2-regular it follows from (exercise 8.4 in [JK]) that it is also simple over $\mathrm{GL}_{\lambda_1'} \, \mathbb{F}_2$.

Consider now an indecomposable reduced injective unstable $\mathcal{A}$-module E. Assume that it is the injective hull of the following simple object in $\mathcal{U}/\mathcal{N}il$:

$$S_\lambda = \bar{C}_\lambda \bar{R}_\lambda \bar{C}_\lambda F(1)^{\otimes n} \cong \bar{R}_{\lambda'} \bar{C}_{\lambda'} \bar{R}_{\lambda'} F(1)^{\otimes n}.$$

We determine, using the method of Chapter 4, the associated simple representation of the corresponding general linear group. The first point is to compute the lowest dimension d such that S_λ embeds in $H^*(\mathbb{Z}/2^{\oplus d})$. Recall that it is also the lowest dimension such that E is a direct summand in $H^*(\mathbb{Z}/2^{\oplus d})$. This is done as follows, as S_λ is simple in $\mathcal{U}/\mathcal{N}il$ it is enough to know that $\mathrm{Hom}_{\mathcal{U}}(S_\lambda, H^*(\mathbb{Z}/2^{\oplus d}))$ is non-trivial, but that $\mathrm{Hom}_{\mathcal{U}}(S_\lambda, H^*(\mathbb{Z}/2^{\oplus d'}))$ is trivial if $d' < d$. Then $\mathrm{Hom}_{\mathcal{U}}(S_\lambda, H^*(\mathbb{Z}/2^{\oplus d})) \cong (T_V(S_\lambda)^0)^* \cong \bar{R}_{\lambda'} \bar{C}_{\lambda'} \bar{R}_{\lambda'} (\mathbb{F}_2^{\oplus d})^{\otimes n}$ and it follows from the combinatorial information contained in [JK] (see also [FS]) that the value we are looking for is λ_1'. Recall the notation $R_d(E)$ of Chapter 4. It remains to determine the representation $R_{\lambda_1'}(E)$ (as an $\mathrm{GL}_{\lambda_1'} \, \mathbb{F}_2$-representation). We have to show that it is the injective hull of $\bar{R}_{\lambda'} \bar{C}_{\lambda'} \bar{R}_{\lambda'} (\mathbb{F}_2^{\oplus \lambda_1'})^{\otimes n}$. As $R_{\lambda_1'}(E)$ is projective and indecomposable, and as $\bar{R}_{\lambda'} \bar{C}_{\lambda'} \bar{R}_{\lambda'} (\mathbb{F}_2^{\oplus \lambda_1'})^{\otimes n}$ is simple, it is enough to show that there is a non-trivial map from $\bar{R}_{\lambda'} \bar{C}_{\lambda'} \bar{R}_{\lambda'} (\mathbb{F}_2^{\oplus \lambda_1'})^{\otimes n}$ to $R_{\lambda_1'}(E)$. But the map $S_\lambda \to E$ induces such a map and we are done.

5.9. The case of an odd prime in the preceding sections

The case of an odd prime is treated in the same way as the case of the prime 2 by using the equivalence of $\mathcal{U}'/\mathcal{N}il'$ (evenly graded unstable $\mathcal{A}$-modules modulo evenly graded nilpotent modules) with $\mathcal{U}/\mathcal{N}il$.

The unstable $\mathcal{A}$-module $F'(2)$ plays the same role as $F(1)$. The polynomial part of K_*^* plays the same role as K_*^* (for $p = 2$).

5.10. The Grothendieck ring of $\mathcal{U}$

Let $G_0(\mathcal{U})$ be the quotient of the free abelian ring generated by isomorphism classes of noetherian unstable $\mathcal{A}$-modules by the ideal generated by elements $[E] - [E'] - [E'']$, for any short exact sequence $0 \to E' \to E \to E'' \to 0$. For any $E \in G_0(\mathcal{U})$ denote by $P_E(q)$ the formal Poincaré series of E. One has (see [Sc2]):

Theorem 5.10.1. *The morphism of rings* $G_0(\mathcal{U}) \to \mathbb{Z}[[q]]$, *sending* $[E]$ *to* $P_E(q)$, *is injective.*

Further, the image is described in [Sc2].

In [CK1] Carlisle and Kuhn have described the Grothendieck ring of finitely generated $\mathbb{F}_p[\text{End } V]$-modules. Considering an adequate colimit process, and taking into account certain torsion conditions, it yields a computation of the Grothendieck ring of the category of noetherian objects in $\mathcal{F}_\omega$. Thus, using the identification between $\mathcal{U}/\mathcal{N}il$ and $\mathcal{F}_\omega$, it yields a computation of the Grothendieck ring of the category of noetherian objects in $\mathcal{U}/\mathcal{N}il$. One gets

Theorem 5.10.2. *The Grothendieck ring of the category of noetherian objects in* $\mathcal{U}/\mathcal{N}il$ *is isomorphic to the quotient of the universal* λ-*ring on one generator quotiented by the ideal generated by the relations* $\psi^p(x) = x$ *for all* x.

In [Sc2] an alternative description of $G_0(\mathcal{U}/\mathcal{N}il)$ is given in terms of certain characters. The description makes use of an Hopf algebra structure on $G_0(\mathcal{U}/\mathcal{N}il)$.

Kuhn also has considered the Grothendieck ring of reduced injective objects of the category $\mathcal{U}$ ([K1]). It is dual to the preceding object (as Hopf algebra). Certain formulas concerning the products can be found in [CK2].

6. Subcategories of $\mathcal{U}$

The object of this chapter is to introduce certain full subcategories of $\mathcal{U}$ and to give some equivalent definitions of them. Then, we study some consequences about the structure of the category $\mathcal{U}$, and we give some applications.

6.1. The categories $\mathcal{N}il_\ell$ and the quotient categories $\mathcal{N}il_\ell/\mathcal{N}il_{\ell+1}$

Let M be an unstable $\mathcal{A}$-module.

Definition-Proposition 6.1.1 [Sc1]. *The unstable $\mathcal{A}$-module M is said to be ℓ-nilpotent, $\ell \geq 0$, if and only if one of the following equivalent conditions holds:*

(i) *M is colimit of unstable modules having a finite filtration whose quotients are ℓ-suspensions;*

(ii) *$\Omega^k M$ is nilpotent for any k such that $0 \leq k \leq \ell - 1$;*

(iii) *$\mathrm{Hom}_{\mathcal{U}}(M, H^*V \otimes N) = 0$ for any V and any N such that $N^j = 0$ for $j > \ell - 1$;*

(iv) *$\mathrm{Hom}_{\mathcal{U}}(M, H^*V \otimes J(n)) = 0$ for any V and any n such that $0 \leq n \leq \ell - 1$;*

(v) *$T_V M$ is $\ell - 1$-connected (i.e. $(T_V M)^j = 0$ if $j \leq \ell$) for any V.*

One notes immediately that (v) is a restatement of (iv) in terms of T_V and that (iii) and (iv) are equivalent. Indeed (iv) $\Rightarrow$ (iii) follows from the fact that an unstable $\mathcal{A}$-module N such that $N^j = 0$ for $j > \ell - 1$ embeds in a product of $J(n)'s$ with $n \leq \ell - 1$. Recall that nilpotent modules have been defined in Section 2.6, Section 1. We shall see that 1-nilpotent and nilpotent mean the same thing. Note that the proposition is empty if $\ell = 0$.

Proof.

(i) $\Rightarrow$ (iv) It is enough to check that $\mathrm{Hom}_{\mathcal{U}}(M, H^*V \otimes J(n)) = 0$, $0 \leq n \leq \ell - 1$, if $M = \Sigma^\ell N$ for an unstable $\mathcal{A}$-module N. But $\mathrm{Hom}_{\mathcal{U}}(\Sigma^\ell N, H^*V \otimes J(n)) \cong \mathrm{Hom}_{\mathcal{U}}(\Sigma^{\ell-n} N, \widetilde{\Sigma}^n(H^*V \otimes J(n)))$ which identifies with $\mathrm{Hom}_{\mathcal{U}}(\Sigma^{\ell-n} N, H^*V)$ by (3.3.5) and it is zero as $\ell - 1 \geq n$ by (2.6.5).

(iii) $\Rightarrow$ (ii) One takes $N = \Sigma^k \mathbb{F}_p$ and by definition, one gets $\mathrm{Hom}_{\mathcal{U}}(\Omega^k M, H^*V) = 0$ for any V. At this point, we use the results of Sections 3.11 and 3.14. Let H be the largest nilpotent module contained in $\Omega^k M$. We have an exact sequence

$$0 \to H \to \Omega^k M \to (\Omega^k M)/H \to 0,$$

and $\Omega^k M/H$ is reduced. If this module is non-trivial, by (3.14.3), there must exist a non-trivial map to some H^*V. This is impossible because
$\mathrm{Hom}_{\mathcal{U}}(\Omega^k M, H^*V) = 0$.

(ii) $\Rightarrow$ (i) One first proves the result for a 0-nilpotent module (thereby identifying 1-nilpotent with nilpotent). This follows from:

— any non-trivial nilpotent module contains a non-trivial suspension,

— a colimit of nilpotent modules is again nilpotent,

— $\mathcal{U}$ is locally noetherian.

Let us do the general case ($\ell \geq 1$). We can assume that M is a suspension and, by induction, that ΩM has a finite filtration whose quotients are $\ell - 1$-suspensions (use a colimit argument). Then, one reduces to the case where ΩM is an $\ell - 1$-suspension. In this case M is an ℓ-suspension. This completes the proof of the theorem.

There is also a characterization of ℓ-nilpotent modules in terms of Steenrod operations [Sc1][HLS3].

If $p = 2$, the operations

$$\mathrm{Sq}_k : M^m \to M^{2m-k}, \quad x \mapsto \mathrm{Sq}^{m-k} x,$$

have to be locally nilpotent for all $0 \leq k < \ell$; if p is an odd prime we write $k = 2\ell + e$ with $e \in \{0, 1\}$ and replace Sq_k by

$$P_k : M^{2m+e} \to M^{2mp+e+2\ell(p-1)}, \quad x \mapsto P^{m-\ell}x .$$

We leave that as an exercise to the reader. We just note that the crucial point is that

$$\Sigma^\ell P_k x = P_{k+\ell} \Sigma^\ell x$$

for any x in an unstable module.

We define $\mathcal{N}il_\ell$ to be the full subcategory of $\mathcal{U}$ that contains all ℓ-nilpotent modules. The categories $\mathcal{N}il_\ell$ are obviously Serre classes in $\mathcal{U}$. It is also clear that $\mathcal{N}il_{\ell+1}$ is contained in $\mathcal{N}il_\ell$ and is a Serre class in $\mathcal{N}il_\ell$. Thus we can consider the quotient categories $\mathcal{N}il_\ell/\mathcal{N}il_{\ell+1}$.

The functor $\Sigma : \mathcal{U} \to \mathcal{U}$ restricts to a functor from $\mathcal{N}il_\ell$ to $\mathcal{N}il_{\ell+1}$ which we shall denote again by Σ. The functor $\Omega : \mathcal{U} \to \mathcal{U}$ restricts to a functor from $\mathcal{N}il_{\ell+1}$ to $\mathcal{N}il_\ell$ by (6.1.1(ii)). We shall denote again this functor by Ω.

Theorem 6.1.2. *The functors Σ and Ω induce mutually inverse equivalences of categories:*

$$\mathcal{N}il_\ell/\mathcal{N}il_{\ell+1} \underset{\Omega}{\overset{\Sigma}{\rightleftarrows}} \mathcal{N}il_{\ell+1}/\mathcal{N}il_{\ell+2} ,$$

for any $\ell \geq 0$.

By the universal property of the quotient categories, Σ induces a functor

$$\mathcal{N}il_\ell/\mathcal{N}il_{\ell+1} \to \mathcal{N}il_{\ell+1}/\mathcal{N}il_{\ell+2} .$$

In order to show that Ω induces a functor

$$\mathcal{N}il_{\ell+1}/\mathcal{N}il_{\ell+2} \to \mathcal{N}il_\ell/\mathcal{N}il_{\ell+1} ,$$

we need to know that the composite functor

$$\mathcal{N}il_{\ell+1} \xrightarrow{\Omega} \mathcal{N}il_\ell \xrightarrow{\mathrm{forget}} \mathcal{N}il_\ell/\mathcal{N}il_{\ell+1}$$

is exact. For that, we use (1.7.5) and we observe that it is enough to show:

Lemma 6.1.3. *If the unstable $\mathcal{A}$-module M is in Nil_ℓ, $\ell \geq 1$ (resp. $\ell = 0$), then $\Omega_1 M$ is in $\mathrm{Nil}_{\ell p - 1}$ (resp. Nil_1).*

Proof. It is enough to consider the case where M is an ℓ-suspension $\Sigma^\ell N$. Recall that $\Sigma \Omega_1 \Sigma^\ell N$ is a submodule of $\Phi \Sigma^\ell N$.

If $p = 2$, $\Phi \Sigma^\ell N$ is isomorphic to $\Sigma^{2\ell} \Phi N$ and we are done if $\ell \geq 1$. If $\ell = 0$ observe that $\Omega_1 N$ is trivial in even degrees and thus a suspension.

If $p > 2$, as Φ does not commute with tensor product, one has to work a bit more. Recall from (1.7.2) that one has the formulas, for any $x \in M$:

$$P^i \Phi x = \Phi P^{i/p} x \,,$$
$$P^i \Phi x = \Phi \beta P^{i-1/p} x \quad \text{if } |x| \equiv 1 \ (2) \,,$$
$$\beta \Phi x = 0 \,.$$

Now let M be $\Sigma^\ell N$. We show that $\Phi \Sigma^\ell N$ is an ℓp-suspension. As there is no action of β on ΦM, it is enough to show that

$$2i > |\Phi x| - \ell p \Rightarrow P^i \Phi x = 0 \,.$$

Assume for example that $\ell \equiv 0$ (2). Write x as $\Sigma^\ell y$ for $y \in N$. There are two cases:

If $|y| \equiv 0$ (2), then $|x| \equiv 0$ (2) and $|\Phi x| = (\ell + |y|)p$. Therefore our condition reads as $2i > p|y|$. But $P^i \Phi x = \Phi P^{i/p} x = \Phi \Sigma^\ell P^{i/p} y$ and the last term is zero because $P^{i/p} y = 0$ as $2\frac{i}{p} > |y|$.

If $|y| \equiv 1$ (2), then $|x| \equiv 1$ (2) and $|\Phi x| = (\ell - 1 + |y|)p + 2$ and our condition reads as $2i > (-1 + |y|)p + 2$ or $2(i-1)/p + 1 > |y|$. But $P^i \Phi x = \Phi \beta P^{i-1/p} x = \Phi \Sigma^\ell \beta P^{i-1/p} y$ and $\beta P^{i-1/p} y = 0$ as $1 + 2(i - 1)/p > |y|$.

The case $\ell \equiv 1$ (2) is left to the reader.

Note that we have proved the following implications:

$$M \in \mathrm{Nil}_\ell \Rightarrow \Phi M \in \mathrm{Nil}_{2\ell} \quad \text{if } p = 2 \,;$$
$$M \in \mathrm{Nil}_\ell \Rightarrow \Phi M \in \mathrm{Nil}_{\ell p} \quad \text{if } p > 2 \,.$$

One concludes that, if $M \in \mathrm{Nil}_\ell$, $\ell \geq 1$, the map $M \twoheadrightarrow \Sigma \Omega M$ is an isomorphism in $\mathrm{Nil}_\ell / \mathrm{Nil}_{\ell+1}$, for its kernel (in $\mathcal{U}$) is in $\mathrm{Nil}_{p\ell}$.

This proves Theorem 6.1.2.

It is clear that the intersection $\bigcap_\ell \mathcal{N}il_\ell$ 'is' the trivial module. The preceding theorem shows that the category $\mathcal{U}/\mathcal{N}il = \mathcal{N}il_0/\mathcal{N}il_1$ is a 'building block' for the category $\mathcal{U}$.

Let M be an unstable $\mathcal{A}$-module. Let N_ℓ be the largest submodule of M which is in $\mathcal{N}il_\ell$. This defines a decreasing filtration on M: $M = N_0 \dots \supset N_\ell \supset N_{\ell+1}$. Here are the two main properties of this filtration (I am indebted to Kuhn to insist that this should be included in this notes, and, indeed, it is worth to note that H.-W. Henn has used a similar property in a crucial way in **[H2]**).

Lemma 6.1.4. *The quotient $N_\ell/N_{\ell+1}$ is the ℓ-th suspension of a reduced module. Moreover if M is finitely generated (as an $\mathcal{A}$-module) this filtration is finite.*

The first part follows easily from the following claim.

Consider an unstable module M such that

(i) M is in $\mathcal{N}il_\ell$,

(ii) any submodule of M which is in $\mathcal{N}il_{\ell+1}$ is trivial.

Then M is the ℓ-th suspension of a reduced module. To prove the claim consider the exact sequence

$$\Phi M \to M \to \Sigma\Omega M \to 0 \,.$$

The arguments used above show and the second condition show that, if $\ell \geq 1$, the first map is trivial. Thus the second one is an isomorphism. Clearly $\Sigma\Omega M$ is in $\mathcal{N}il_{\ell-1}$, and does contain non-trivial submodule which are in $\mathcal{N}il_\ell$. One concludes by induction.

The second part follows from 3.11. As M is a finitely generated unstable $\mathcal{A}$-module it embeds in a finite direct sum of $H^*V \otimes J(n)$'s. Then one observes that $H^*V \otimes J(n)$ does not contain non-trivial $n + 1$-th suspension, hence does not contain a non-trivial submodule which is in $\mathcal{N}il_{n+1}$. The result follows.

6.2. The category $\mathcal{B}$ of locally finite unstable $\mathcal{A}$-modules and the categories $\widetilde{Nil}_\ell$

In this section, we introduce some other full subcategories of $\mathcal{U}$ which are important for applications.

Recall that an $\mathcal{A}$-module M is said to be locally finite if and only if, for any x in M, the span of x over $\mathcal{A}$ is finite (as a set or as an $\mathbb{F}_p$-vector space!).

Denote by $\mathcal{B}$ the full subcategory of $\mathcal{U}$ whose objects are the unstable modules which are locally finite.

Theorem 6.2.1 (see [LS2]). *Let M be an unstable $\mathcal{A}$-module. The following conditions are equivalent:*

(i)   *M is locally finite;*

(ii)   *$\mathrm{Hom}_{\mathcal{U}}(M, \widetilde{H}^*V \otimes J(m))$ is trivial for any V and $m \geq 0$;*

(iii)   *$T_V M$ is isomorphic to M for any V;*

(iv)   *$T M$ is isomorphic to M;*

(v)   *$\overline{T}_V M$ is trivial for any V;*

(vi)   *$\overline{T} M$ is trivial.*

Proof.

(i) $\Rightarrow$ **(ii)** One shows that, for any non-zero $x \in \widetilde{H}^*V \otimes J(m)$, the span of x over $\mathcal{A}$ is infinite. But, if one filters $J(m)$ by degree, one gets a (finite) filtration of $\widetilde{H}^*V \otimes J(m)$ whose quotients are suspensions of $\widetilde{H}^*V$. Therefore it is enough to show that, for any non-zero $x \in \widetilde{H}^*V$, $\mathcal{A}x$ is infinite. If this is not true, $\widetilde{H}^*V$ contains a non-zero element y of positive degree such that $\mathcal{A}y = \mathbb{F}_p y$. Therefore $\widetilde{H}^*V$ would contain a non trivial suspension, and this is false by (2.6.5).

(ii) $\Rightarrow$ **(i)** This is a consequence of 3.11. It is sufficient to show that, if an $\mathcal{A}$-cyclic module $\mathcal{A}x$ is infinite, $\mathrm{Hom}_{\mathcal{U}}(\mathcal{A}x, \widetilde{H}^*V \otimes J(m))$ is non trivial for some V and some $m \geq 0$. This is done as follows.

An noetherian unstable $\mathcal{A}$-module M (i.e. an M having finitely many generators over $\mathcal{A}$) has an injective hull which is the direct sum of finitely many indecomposable $\mathcal{U}$-injectives. Therefore as $\mathcal{A}x$ is infinite, its injective hull cannot be a direct sum of $J(n)$'s, and there is a non trivial map from $\mathcal{A}x$ to $L \otimes J(m)$ for some $L \in \mathcal{L}_d$, $d \geq 1$, $m \geq 0$.

Finally, (iii) is a restatement of (ii) using T_V, (iv) is equivalent to (iii), and (v) and (vi) are restatement of (iii) and (iv) using $\bar{T}_V$ and $\bar{T}$.

Denote by $\overline{Nil}_\ell$ the smallest (abelian) full subcategory of $\mathcal{U}$ which contains Nil_ℓ and $\mathcal{B}$, which is a Serre class in $\mathcal{U}$, and which is stable under colimits.

Theorem 6.2.2. *Let M be an unstable $\mathcal{A}$-module. The following conditions are equivalent, let $\ell \geq 1$*

(i) M *is in* $\overline{Nil}_\ell$;

(ii) $\Omega^k M$ *is in* $\overline{Nil}_1$ *for any k such that $0 \leq k \leq \ell - 1$;*

(iii) $\mathrm{Hom}_{\mathcal{U}}(M, \tilde{H}^*V \otimes J(m))$ *is trivial for any V and m, such that $0 \leq m \leq \ell - 1$.*

This is left to the reader.

For a concept analogous to the one of locally finite $\mathcal{A}$-module in the context of right $\mathcal{A}$-modules, we refer to **[LS1]**.

The category $\mathcal{B}$ is the first step of the Krull filtration of the category $\mathcal{U}$. Recall the definition of this filtration **[Gb]**. The full subcategory $\mathcal{U}_0$ is the smallest Serre class in $\mathcal{U}$ which is stable under colimits and contains all simple objects in $\mathcal{U}$. As any simple objects in $\mathcal{U}$ is of the form $\Sigma^n \mathbb{F}_p$, one gets $\mathcal{U}_0 = \mathcal{B}$. Suppose $\mathcal{U}_n$ has been defined. Define $\mathcal{U}_{n+1}$ as follows. Consider in the quotient category $\mathcal{U}/\mathcal{U}_n$ the smallest Serre class which is stable under colimits and contains all simple objects (of $\mathcal{U}/\mathcal{U}_n$). Then an object M in $\mathcal{U}$ belongs to $\mathcal{U}_n$, if and only if, as an object of $\mathcal{U}/\mathcal{U}_n$, it belongs to the subcategory we have just described.

Theorem 6.2.3. *The smallest abelian subcategory of $\mathcal{U}$ which contains all $\mathcal{U}_n$, $n \in \mathbb{N}$, and is stable under colimits, is $\mathcal{U}$ itself.*

This will follow from Theorem 6.2.4. Note that within the context of a general abelian category $\mathcal{C}$ the analogous result is not true. One needs to introduce sub-categories $\mathcal{C}_\alpha$ indexed over the ordinals. Then, in order to have a statement of the type of (6.2.3), one needs to know that $\mathcal{C}$ is locally noetherian [**Gb**]. It is possible to characterize the subcategories $\mathcal{U}_n$ in another way.

Theorem 6.2.4. *Let M be an unstable A-module. Then the following two conditions are equivalent:*

(i) M *is in* $\mathcal{U}_n$*;*

(ii) $\mathrm{Hom}_{\mathcal{U}}(M, (\widetilde{H}^* \mathbb{Z}/p)^{\otimes n+1} \otimes J(m))$ *is trivial for any* m*;*

(iii) $\overline{T}^{n+1} M$ *is trivial.*

We shall only outline the proof. It depends on the following characterization of unstable modules M such that $\overline{T}^{n+1} M = 0$ whose proof follows from 5.5.7 and is left to the reader.

Proposition 6.2.5. *Let M be an unstable module. Then $\overline{T}^{n+1} M = 0$ if and only if M is in the smallest full subcategory $\mathcal{U}'_n$ of $\mathcal{U}$ defined by the two following conditions*

(i) $\mathcal{U}'_n$ *is stable under colimits and suspensions, and*

(ii) $\mathcal{U}'_n$ *contains the images by the functor s_1 of the objects of the subcategory $\mathcal{V}_n$ of $\mathcal{U}/\mathcal{N}il$.*

Proof of 6.2.4. It remains to identify $\mathcal{U}_n$ with $\mathcal{U}'_n$. This is done by induction on n. The case $n = 0$ is easy. Then it is enough to identify simple objects in the quotient category $\mathcal{U}/\mathcal{U}_{n-1}$.

Proposition 6.2.6. *A set of representatives of simple objects in the category $\mathcal{U}/\mathcal{U}_{n-1}$ is given by the objects $\Sigma^\ell s_1 R$, $\ell \geq 0$, and R runs through a set of representatives of simple objects in $\mathcal{V}_n - \mathcal{V}_{n-1}$.*

It is clear that these are simple objects in the category $\mathcal{U}/\mathcal{U}_{n-1}$. That these are the only one up to isomorphism is shown as follows. One first observes, by using 6.1.4, that any simple object is going to be the isomorphism class of an iterated suspension of a reduced module R. Then one notes that R in $\mathcal{U}/\mathcal{N}il$ represents the class of a simple object which is in $\mathcal{V}_n - \mathcal{V}_{n-1}$. Indeed, using 1.7.4, one shows that an object of the form $s_1 R$, $R \in \mathcal{V}_\ell - \mathcal{V}_{\ell-1}$, $\ell > n$ is not simple in $\mathcal{U}/\mathcal{U}_{n-1}$.

6.3. Localization away from $\mathcal{N}il_\ell$

The forgetful functor $r_\ell : \mathcal{U} \to \mathcal{U}/\mathcal{N}il_\ell$ has a right adjoint s_ℓ. This follows from ([**Gb**], chap.3, 3).

Proposition 6.3.1. *Let $\mathcal{D}$ be a Serre class in an abelian category $\mathcal{C}$. Assume that injective hulls exist in $\mathcal{C}$. Then the following conditions are equivalent:*

(i) *the forgetful functor $\mathcal{C} \to \mathcal{C}/\mathcal{D}$ has a right adjoint;*

(ii) *any object M in $\mathcal{C}$ contains a subobject which is maximal among subobjects of $\mathcal{C}$ belonging to $\mathcal{D}$.*

The condition (ii) is satisfied for $\mathcal{U}$ and $\mathcal{N}il_\ell$.

The unit of the adjunction $M \to s_\ell \circ r_\ell M$ should be thought of as a localization away from $\mathcal{N}il_\ell$. The unstable $\mathcal{A}$-module M is said to be $\mathcal{N}il_\ell$-closed if this map is an isomorphism. We are going to give a criterion ensuring that an object in $\mathcal{U}$ is $\mathcal{N}il_\ell$-closed. More generally, consider the situation of a general abelian category $\mathcal{C}$ and of a Serre class $\mathcal{D}$. Assume that the forgetful functor $r : \mathcal{C} \to \mathcal{C}/\mathcal{D}$ has a right adjoint s. An object $C \in \mathcal{C}$ is said to be $\mathcal{D}$-closed if the unit of the adjunction $C \to s_1 \circ r_1 C$ is an isomorphism. The following is not hard to prove [**Gb**]:

Proposition 6.3.2. *Let C be an object of $\mathcal{C}$. The following two conditions are equivalent:*

(i) *C is $\mathcal{D}$-closed;*

(ii) *$\mathrm{Ext}^i_{\mathcal{C}}(D, C) = 0$ for any $D \in \mathcal{D}$ and $i = 0, 1$.*

In our case, these conditions are

(i) *M is $\mathcal{N}il_\ell$-closed;*

(ii) *$\mathrm{Ext}^i_{\mathcal{U}}(N, M) = 0$ for any $N \in \mathcal{N}il_\ell$ and $i = 0, 1$.*

The case where $\ell = 1$ is of special importance. It has been studied in detail by Lannes and Zarati [**LZ1**]. They have given the following description of the localization functor $\ell_1 = s_1 \circ r_1$.

Theorem 6.3.3. *The functors $M \mapsto \mathrm{colim}_k \widetilde{\Phi}^k M$ and ℓ_1 are naturally equivalent.*

This is implied by the following:

Theorem 6.3.4. *Let M be an unstable A-module. Then the following conditions are equivalent:*

(i) *M is Nil-closed;*

(ii) $\widetilde{\lambda}: M \to \widetilde{\Phi}M$ *is an isomorphism;*

(iii) *M is reduced and the quotient E_M/M of the injective hull E_M of M by M is reduced;*

(iv) *for $p = 2$, M is reduced and an element x of M is of the form $\mathrm{Sq}_0\, y$ for some $y \in M$ if and only if $Q_i x = 0$ for all Milnor operations Q_i.*

Proof of Theorem 6.3.4 **[LZ1]**.

(i) $\Rightarrow$ (ii) follows easily from

Lemma 6.3.5. *For any unstable A-module M, there is an exact sequence*

$$0 \to \Sigma\widetilde{\Sigma}M \to M \xrightarrow{\widetilde{\lambda}} \widetilde{\Phi}M \to \Sigma R^1\widetilde{\Sigma}M \to 0 \ ,$$

where $R^1\widetilde{\Sigma}$ is the first right derived functor of $\widetilde{\Sigma}$, and $\Sigma\widetilde{\Sigma}M \to M$ is the counit of the adjunction.

Proof. One applies the functor $\mathrm{Hom}_{\mathcal{U}}(-, M)$ to the exact sequence $0 \to \Phi F(m) \to F(m) \to \Sigma F(m-1) \to 0$ and one gets an exact sequence:

$$0 \to (\Sigma\widetilde{\Sigma}M)^m \to M^m \to (\widetilde{\Phi}M)^m \to \mathrm{Ext}^1_{\mathcal{U}}(\Sigma F(m-1), M) \to 0.$$

Then it is easy to identify $\mathrm{Ext}^1_{\mathcal{U}}(\Sigma F(m-1), M)$ with $\Sigma R^1\widetilde{\Sigma}M$ in degree m. However, the only point that we really need, to prove that (i) $\Rightarrow$ (ii), is that coker $\widetilde{\lambda}$ is a suspension, which is immediate.

(ii) $\Rightarrow$ (i) follows also from Lemma 6.3.5, for it shows that M is reduced. It remains to show that $\mathrm{Ext}^1_{\mathcal{U}}(N, M) = 0$ for any N in Nil. If this is not the case one checks there exists an object $\overline{M}$ in $\mathcal{U}$ such that:

(i) M is included in, but different of, $\overline{M}$ and $\overline{M}$ is reduced, and

(ii) $\Phi\overline{M} \subset M$.

To find $\overline{M}$, one observes that if $\mathrm{Ext}^1_{\mathcal{U}}(N, M)$ is non-trivial for some $N \in \mathcal{N}il$, it is non-trivial for a certain suspension. One observes that this suspension can be supposed to be cyclic as an $\mathcal{A}$-module (and moreover, if $p > 2$, to have a trivial action of β).

Then one has the following sequence of maps

$$\Phi M \hookrightarrow \Phi \overline{M} \hookrightarrow M ,$$

which yields by adjunction a factorization of $\widetilde{\lambda}$:

$$M \hookrightarrow \overline{M} \to \widetilde{\Phi} M$$

which shows that $\widetilde{\lambda}$ is not an isomorphism. The result follows.

(ii) $\Rightarrow$ (iii) is left to the reader, as well as (iii) $\Rightarrow$ (ii).

It remains to look at condition (iv); for that we refer to [LZ1]. We note that it depends on the following exact sequence, which is the beginning of a projective resolution of $\Sigma F(n-1)$:

$$\oplus_{i \geq 1} F(2n+2^i - 1) \xrightarrow{d_1} F(2n) \xrightarrow{d_0} F(n) \xrightarrow{d_{-1}} \Sigma F(n-1) \to 0 ,$$

where the maps are defined by

$$d_{-1}(i_n) = \Sigma i_{n-1}, \quad d_0 i_{2n} = \mathrm{Sq}_0 \, i_n, \quad d_1 i_{2n+2^i - 1} = Q_{i-1} i_{2n} .$$

We note to conclude this section that, in [BZ1] and [BZ2], Broto and Zarati have analyzed 'away from $\mathcal{N}il_\ell$'-localization in the context of unstable $\mathcal{A}$-algebras.

6.4. The categories $\mathcal{N}il_\ell$ and the functors Tor

Now we give an application of the preceding results. Let K be an unstable $\mathcal{A}$-algebra and M be an unstable $\mathcal{A}$-module which is also a K-module. Assume that the structure map $m : K \otimes M \to M$ is $\mathcal{A}$-linear. Moreover assume that K is augmented, let $\varepsilon : K \to \mathbb{F}_p$ be the augmentation. With these assumptions the graded $\mathbb{F}_p$-vector space $\mathrm{Tor}^s_K(M, \mathbb{F}_p)$, $s \leq 0$ (as usual for Tor with upper indices), is in a natural way an unstable $\mathcal{A}$-module. Indeed the defining complex

$$\ldots \to M \otimes \overline{K}^{\otimes -s} \to M \otimes \overline{K}^{\otimes(-s-1)} \to \ldots$$

is a complex of unstable $\mathcal{A}$-modules ($\overline{K}$ is the kernel of the augmentation).

Theorem 6.4.1. *The unstable $\mathcal{A}$-module $\operatorname{Tor}_K^s(M, \mathbb{F}_p)$ is in the category $\mathcal{N}il_{-s}$.*

Later we shall discuss applications of this theorem to the Eilenberg-Moore spectral sequence. Observe that this theorem is a generalization of the following well-known facts:

— if $p = 2$, $\operatorname{Tor}_K^{-1}(\mathbb{F}_2, \mathbb{F}_2) \cong \bar{K}/\bar{K} \cdot \bar{K}$ is a suspension,

— if $p > 2$, $\operatorname{Tor}_K^{-1}(\mathbb{F}_2, \mathbb{F}_2) \cong \bar{K}/\bar{K} \cdot \bar{K}$ is nilpotent.

Proof. We shall apply the following criterion of Proposition 6.1.1 (v): $M \in \mathcal{N}il_\ell \Leftrightarrow (T_V M)^j = 0$ for all V and all j such that $0 \le j \le \ell - 1$. In other words, one has to check that $T_V M$ is $\ell - 1$-connected for all V.

We are going to apply this criteria.

Lemma 6.4.2. *There is an isomorphism*

$$T_V(\operatorname{Tor}_K^s(M, \mathbb{F}_p)) \cong \operatorname{Tor}_{T_V(K)}^s(T_V(M), \mathbb{F}_p).$$

The $T_V(K)$-module structure on $T_V(M)$ is given by

$$T_V(K) \otimes T_V(M) \cong T_V(K \otimes M) \xrightarrow{T_V(m)} T_V(M).$$

This map is $\mathcal{A}$-linear. The augmentation $T_V(K) \to \mathbb{F}_p$ is just $T_V(\varepsilon)$. The lemma follows from the exactness of T_V, and from the commutation of T_V with tensor products. Indeed, when one applies T_V to the defining complex of $\operatorname{Tor}_K^*(M, \mathbb{F}_p)$, one gets the defining complex for $\operatorname{Tor}_{T_V(K)}^*(T_V(M), \mathbb{F}_p)$.

In view of the lemma, it is enough to show that for any K and any M, the unstable module $\operatorname{Tor}_K^s(M, \mathbb{F}_p)$ is $(-s - 1)$-connected. This is true by routine argument if K is connected (i.e. if $K^0 = \mathbb{F}_p$). But, in our case, K need not be connected. We are going to use the fact that K^0 is a p-Boolean algebra. Consider a more general situation:

Let L be a unital, commutative $\mathbb{N}$-graded $\mathbb{F}_p$-algebra which:

(i) has an augmentation $\varepsilon : L^0 \to \mathbb{F}_p$;

(ii) and is such that L^0 is a p-Boolean algebra.

Let M be an ($\mathbb{N}$-graded) L-module.

Lemma 6.4.3. *With the preceding assumptions, the* $\mathbb{N}$-*graded* $\mathbb{F}_p$-*vector space* $\mathrm{Tor}_L^s(L, \mathbb{F}_p)$ *is* $(-s-1)$-*connected.*

Proof. Let L_ε be the algebra $L \otimes_{L^0} \mathbb{F}_p$ ($\mathbb{F}_p$ is an L^0-module via ε). Then L_ε is connected. Let M_ε be the L_ε-module $M \otimes_{L^0} \mathbb{F}_p$. We claim that:

$$\mathrm{Tor}_L^s(M, \mathbb{F}_p) \cong \mathrm{Tor}_{L_\varepsilon}^s(M_\varepsilon, \mathbb{F}_p).$$

This follows from the change of rings spectral sequence [CE];

$$\mathrm{Tor}_{L_\varepsilon}^p(\mathrm{Tor}_L^q(M, L_\varepsilon), \mathbb{F}_p) \Rightarrow \mathrm{Tor}_L^{p+q}(M, \mathbb{F}_p);$$

there $\mathbb{F}_p$ is a projective L^0-module (L^0 being p-Boolean), so L_ε is an projective L-module and the spectral sequence collapses to the indicated isomorphism.

Remark. Theorem 6.4.1 extends to the case of $\mathrm{Tor}_K^s(M, N)$.

The proof is nearly the same as above. We have to show that $\mathrm{Tor}_K^s(M, N)$ is $(-s-1)$-connected for any (non-connected) unstable algebra K, any M and any N. We reduce to the case where K is finite type by a colimit argument. Let $\varepsilon_1, ..., \varepsilon_t$ be all the augmentation homomorphims of K (i.e. ring homomorphisms from K to $\mathbb{F}_p$). Let N_i be $N \otimes_{K^0} \mathbb{F}_p$, where $\mathbb{F}_p$ is provided with the K-module structure given by ε_i. Then, as K-module, N is isomorphic to the direct sum $\oplus_i N_i$. This, because K^0 is a p-Boolean algebra. Therefore, we must consider $\mathrm{Tor}_K^s(M, N_i)$. In this case the change of ring spectral sequence shows that

$$\mathrm{Tor}_K^s(M, N_i) \cong \mathrm{Tor}_{K_i}^s(M_i, N_i),$$

where K_i is defined as was N_i. One completes the proof by proceeding as above.

We conclude this section with an application to unstable $\mathcal{A}$-algebras.

Proposition 6.4.4. *Let* K *be an augmented unstable* $\mathcal{A}$-*algebra. Then, if* $\mathrm{Tor}_K^{-1}(\mathbb{F}_p, \mathbb{F}_p) \cong Q(K)$ *is locally finite as an* $\mathcal{A}$-*module, so is* $\mathrm{Tor}_K^s(\mathbb{F}_p, \mathbb{F}_p)$, $s \leq -1$.

Proof. Theorem 6.2.1 shows that it is enough to show that

$$T_V(\mathrm{Tor}_K^s(\mathbb{F}_p, \mathbb{F}_p)) \cong \mathrm{Tor}_K^s(\mathbb{F}_p, \mathbb{F}_p).$$

But by Lemma 6.4.2 and 6.4.3, abbreviating $T_V(\varepsilon)$ by ε, we have

$$T_V(\mathrm{Tor}^s_K(\mathbb{F}_p, \mathbb{F}_p)) \cong \mathrm{Tor}^s_{T_V(K)}(\mathbb{F}_p, \mathbb{F}_p)$$
$$\cong \mathrm{Tor}^s_{T_V(K)_\varepsilon}(\mathbb{F}_p, \mathbb{F}_p).$$

But it follows from (3.9.7) that if $Q(K)$ is locally finite, $T_V(K)_\varepsilon \cong K$. So we are done.

We note also a converse to (3.9.7) (compare with **[DW2]**):

Proposition 6.4.5. *Let K be an augmented unstable $\mathcal{A}$-algebra. Assume that the component $T_V(K)_{T_V(\varepsilon)}$ of $T_V(K)$ is isomorphic to K for all V. Then $Q(K)$ is locally finite as an unstable $\mathcal{A}$-module.*

Proof.   We know that $Q(T_V(K)_{T_V(\varepsilon)}) \cong T_V Q(K)$. Thus Theorem 6.2.1 gives the result.

PART

3

THREE

The Sullivan conjecture
and the cohomology of mapping spaces

7. Non-Abelian homological algebra and André-Quillen cohomology

The first aim of this chapter is to introduce some basic facts about simplicial resolutions and comonad-derived functors. We shall be concerned with functors from $\mathcal{K}$ or $\mathcal{K}_a$ (augmented unstable $\mathcal{A}$-algebras) to $\mathcal{U}$, $\mathcal{E}$ or $\mathcal{E}_{gr}$ ($\mathbb{N}$-graded $\mathbb{F}_p$-vector spaces), or from $\mathcal{A}lg$ ($\mathbb{N}$-graded, commutative, unital, $\mathbb{F}_p$-algebras) or $\mathcal{A}lg_a$ (augmented objects of $\mathcal{A}lg$) to $\mathcal{E}$ or $\mathcal{E}_{gr}$. In the sequel we shall also consider functors from the category $\mathcal{K}/K$ of unstable $\mathcal{A}$-algebras over an unstable $\mathcal{A}$-algebra K.

In the second part of the chapter we discuss some properties, as $\mathcal{A}$-module, of the André-Quillen cohomology (which is defined there) of unstable $\mathcal{A}$-algebras. Then, we introduce derivations, and prove a theorem of Lannes that is going to be crucial for the homotopical applications.

7.1. Simplicial resolutions in the categories $\mathcal{K}$ and $\mathcal{A}lg$

The category $\mathcal{K}$ is provided with a comonad (G, η, ν) (see Section 3.8). Recall how $G : \mathcal{K} \to \mathcal{K}$ is built up. The forgetful functor $\mathcal{O} : \mathcal{K} \to \mathcal{E}_{gr}$, has a left adjoint $\mathcal{G} : \mathcal{E}_{gr} \to \mathcal{K}$. The functor $\mathcal{G}$ is the composite, of the left adjoint of the forgetful functor $\mathcal{O} : \mathcal{U} \to \mathcal{E}_{gr}$ (denoted $\mathcal{F}$ in Chapter 3), and of the Steenrod-Epstein functor U which is itself left adjoint to the forgetful functor $\mathcal{O} : \mathcal{K} \to \mathcal{U}$. The functor $G : \mathcal{K} \to \mathcal{K}$ is the composite $\mathcal{G} \circ \mathcal{O}$.

155

The category $\mathcal{A}lg$ is provided with a comonad (S, η, ν) where S is the composite functor $\mathcal{S} \circ \mathcal{O}$, where $\mathcal{O} : \mathcal{A}lg \to \mathcal{E}_{gr}$ is as usual the forgetful functor. The functor $\mathcal{S} : \mathcal{E}_{gr} \to \mathcal{A}lg$ is the usual symmetric algebra functor, left adjoint to $\mathcal{O}$.

The categories $\mathcal{K}_a$ and $\mathcal{A}lg_a$ have similarly defined comonads. We shall also denote these comonads by G (resp. S). The context will make it clear whether we are working with augmented algebras or not.

Recall that a simplicial object in a category $\mathcal{C}$ is a contravariant functor from Δ to $\mathcal{C}$ (see Chapter 3). It can also be seen as a graded object $\{C_n\}$, $n \in \mathbb{N}$, provided with morphisms:

$$d_i : C_n \to C_{n-1}, \quad 0 \le i \le n,$$
$$s_j : C_n \to C_{n+1}, \quad 0 \le j \le n$$

satisfying the identities (3.2.4).

In the following definition, $\mathcal{C}$ will denote one of the following four categories: $\mathcal{K}$, $\mathcal{A}lg$, $\mathcal{K}_a$, $\mathcal{A}lg_a$; and (H, η, ν) will denote the comonad on $\mathcal{C}$.

Definition 7.1.1. A simplicial resolution of an object C in $\mathcal{C}$ is a simplicial object $C_\bullet$ in $\mathcal{C}$ such that the following conditions hold:

(i) $C_\bullet$ is augmented over C, *i.e.* there is a morphism $\varepsilon : C_0 \to C$ such that $\varepsilon \circ d_0$ and $\varepsilon \circ d_1 : C_1 \to C$ are equal;

(ii) for all i, there exists D_i in $\mathcal{C}$ such that $C_i \cong H(D_i)$,

(iii) the following complex of graded vector spaces is acyclic:

$$\ldots \mathcal{O}(C_2) \xrightarrow{d_2-d_1+d_0} \mathcal{O}(C_1) \xrightarrow{d_1-d_0} \mathcal{O}(C_0) \to \mathcal{O}(C) \to 0,$$

where $\mathcal{O} : \mathcal{C} \to \mathcal{E}$ is the forgetful functor and, by abuse of notations, $\mathcal{O}(d_i)$ is denoted by d_i.

The 'standard resolutions' of Chapter 3 obviously satisfy (i) and (ii). Let us (re)prove that they satisfy (iii). This follows from the following lemma, where we keep the notations of (7.1.1).

Lemma 7.1.2. *Let E be a functor from $\mathcal{C}$ in an abelian category. Assume there is a natural transformation $h : E \to E \circ H$ such that,* *for all* $C \in \mathcal{C}$, *the composite* $E(\eta_C) \circ h_C : E(C) \to E(C)$ *is the identity. Then for all* $C \in \mathcal{C}$, *the following complex is acyclic*

$$\to E\left(H_{\bullet}(C)_n\right) \xrightarrow{\Sigma(-1)^i d_i} E\left(H_{\bullet}(C)_{n-1}\right) \to \ldots \to E(C) \to 0 .$$

Recall that $H_{\bullet}(C)_n = H^{n+1}(C)$, and that $d_i = H^i(\eta_{H^{n-i}(C)})$, where η is the counit of the adjunction.

Proof.   Define $h_n : E(H_{\bullet}(C)_n) \to E(H_{\bullet}(C)_{n+1})$ by $h_n = h_{H_{\bullet}(C)_n}$, $n \geq 0$; also let h_{-1} denote $h : C \to H(C)$. One checks that:

(i) $d_0 \circ h_n = \text{Id}$, and

(ii) $d_i \circ h_n = h_{n-1} \circ d_{i-1}$, for $i \geq 1$. From this it follows (as E takes values in an abelian category) that:

$$h_{n-1} \circ \left(\sum_0^n (-1)^i d_i \right) + \left(\sum_0^{n+1} (-1)^i d_i \right) \circ h_n = \text{Id} .$$

The lemma follows. The map h is called a contraction.

In the case we are interested in E is the forgetful functor $\mathcal{O}$ taking values in the category $\mathcal{E}_{gr}$ of $\mathbb{N}$-graded $\mathbb{F}_p$-vector spaces. The natural transformation $h : \mathcal{O} \to \mathcal{O} \circ G$ (resp. $h : \mathcal{O} \to \mathcal{O} \circ S$) is the adjoint of the identity $G = \mathcal{G} \circ \mathcal{O} \to G = \mathcal{G} \circ \mathcal{O}$ (resp. $\mathcal{S} \circ \mathcal{O} \to \mathcal{S} \circ \mathcal{O}$). The condition (iii) of Definition 7.1.1 will hold as soon as the comonad H arises from an adjoint pair: $F : \mathcal{C} \to \mathcal{D}$, $G : \mathcal{D} \to \mathcal{C}$ (where G is a left adjoint), $\mathcal{D}$ abelian, and moreover if E factors through F. Indeed the preceding argument shows that the complex $\{F(H_{\bullet}(C)_n), \sum(-1)^i d_i\}$, $n \in \mathbb{N}$ is acyclic. Denote the contraction by h. If E can be written as $F' \circ F$ it is clear that $F'(h)$ is a contraction for the complex $\{E(H_{\bullet}(C)_n), \sum(-i)^i d_i\}$, $n \in \mathbb{N}$.

The following lemma, in which we keep the notations of (7.1.2), justifies the terminology 'H-free' or 'H-projective' (more simply free or projective) for objects of the form $H(D)$ in $\mathcal{C}$.

Lemma 7.1.3. *Let* C *be an object of* $\mathcal{C}$ *and suppose there is a morphism* $r : C \to H(C)$ *such that the composite* $C \to H(C) \to C$

is the identity. Then for any functor E *from* $\mathcal{C}$ *to an abelian category the following complex is acyclic:*

$$\to E\left(H_{\bullet}(C)_n\right) \xrightarrow{\Sigma(-1)^i d_i} E\left(H_{\bullet}(C)_{n-1}\right) \to \ldots \to E(C) \to 0.$$

The main examples we have in mind are $\mathcal{C} = \mathcal{K}, \mathcal{A}lg, \mathcal{K}_a, \mathcal{A}lg_a$ and E is a forgetful or augmentation ideal functor to $\mathcal{U}$ or $\mathcal{E}_{gr}$, where C is of the form $G(K), U(M)$ or $S(A)$ (with $K \in \mathcal{K}, M \in \mathcal{U}$ and $A \in \mathcal{A}lg$). In all of these cases, the morphism r is just $\nu_{G(K)} : G(K) \to G^2(K)$ (resp. $\nu_{S(A)} \ldots$).

Free objects of $\mathcal{A}lg$ *are symmetric algebras in the graded sense. Connected objects of the form* $G(K), U(M)$ *are free as algebras* (see for example **[BK1]** or Chapter 3).

The proof goes as follows: let $r_n : H_{\bullet}(C)_n \to H_{\bullet}(C)_{n+1}$ be $H^{n+1}(r)$ if $n \geq 0$, and let r_{-1} be r. The following identities hold:

(i) $d_i \circ r_n = r_{n-1} \circ d_i$, $n \geq 0$, $0 \leq i \leq n$. For $n = 0$, $r_{n-1} \circ d_i$ is replaced by $r_{n-1} \circ \eta_C$.

(ii) $d_{n+1} \circ r_n = \mathrm{Id}$. The result follows easily.

We now consider the following situation: $\mathcal{C}$ and H are as above, the comonad H is defined by an adjoint pair of functors $\mathcal{H}, \mathcal{O}$, where $\mathcal{O}$ takes values in $\mathcal{E}_{gr}$, and $\mathcal{H}$ is the left adjoint to $\mathcal{O}$. Consider also a functor $E : \mathcal{C} \to \mathcal{D}$, with $\mathcal{D}$ abelian.

Definition 7.1.4 [BB][Bo1]. The i-th derived functor of E, with respect to H, evaluated at $C \in \mathcal{C}$ is the i-th homology object of the complex

$$\left\{E(C_{\bullet})_n, \sum_j (-1)^j E(d_j)\right\}_{n \in \mathbb{N}},$$

$C_{\bullet}$ being any simplicial resolution of C. It is denoted $L_i^H E(C)$.

This definition deserves two comments:

(i) Simplicial resolutions do exist.

(ii) The value of $L_i^H E(C)$ is independent of the choice of the resolution.

Part (i) was shown in (7.1.2).

Part (ii) is shown as follows. For a set S denote by $\mathbb{Z}[S]$ the free abelian group with basis S. Now let $C_{\bullet}$ and $C'_{\bullet}$ be two simplicial resolutions of C. It follows from the proof of Lemma 7.1.2 that the following complex is acyclic for any object $D \in \mathcal{D}$:

$$\left\{ \mathbb{Z}\left[\mathrm{Hom}_{\mathcal{C}}(\mathcal{H}(D), C_{\bullet}) \right]_n , \sum_j (-1)^j \, \mathrm{Hom}_{\mathcal{C}}(\mathcal{H}(D), d_j) \right\}_{n \in \mathbb{N}} ;$$

this, because $\mathrm{Hom}_{\mathcal{C}}(\mathcal{H}(D), C_n) \cong \mathrm{Hom}_{\mathcal{D}}(D, \mathcal{O}(C_n))$. From this point, the standard proof works a usual. Denote by $\varepsilon : C_0 \to C$, and $\varepsilon' : C'_0 \to C'$ the augmentation homomorphisms. One gets elements

$$f_n \in \mathbb{Z}\left[\mathrm{Hom}_{\mathcal{C}}(C_n, C'_n) \right] \quad \text{and} \quad g_n \in \mathbb{Z}\left[\mathrm{Hom}_{\mathcal{C}}(C'_n, C_n) \right]$$

such that

(i) $\varepsilon' \circ f_0 = \varepsilon, \varepsilon \circ g_0 = \varepsilon'$;

(ii) $\left(\sum_i (-1)^i d_i \right) \circ f_n = f_{n-1} \circ \left(\sum_i (-1)^i d_i \right)$;

(iii) $\left(\sum_i (-1)^i d_i \right) \circ g_n = g_{n-1} \circ \left(\sum_i (-1)^i d^i \right)$.

Denote the composition $g_n \circ f_n$ by π_n. One constructs elements h_n in $\mathbb{Z}\left[\mathrm{Hom}_{\mathcal{C}}(C_{n-1}, C_n) \right]$ such that

$$\mathrm{Id} - \pi_n = h_{n-1} \circ \left(\sum_i (-1)^i d_i \right) + \left(\sum_i (-1)^i d_i \right) \circ h_n$$

in $\mathbb{Z}\left[\mathrm{Hom}_{\mathcal{C}}(C_n, C_n) \right]$. An element of $\mathbb{Z}\left[\mathrm{Hom}_{\mathcal{C}}(C, D) \right]$ induces a well defined map from $E(C)$ to $E(D)$ for any functor E from $\mathcal{C}$ to an additive category, and the result follows.

The main examples we have in mind are the following.

7.1.5. $\mathcal{C} = \mathcal{A}lg_a$ and $E : \mathcal{A}lg_a \to \mathcal{E}_{gr}$ is the 'indecomposable functor', $Q(A) = \bar{A}/\bar{A} \cdot \bar{A}$, where $\bar{A}$ is the augmentation ideal. The derived functors, called the André-Quillen cohomology of A, are denoted $L_i^S Q(A)$ and have been introduced by Quillen [**Q1**] and André [**An**]. They are denoted by $D_{i+1}(\mathbb{F}_p/A)$ in these references.

7.1.6. $\mathcal{C} = \mathcal{K}_a$, $E : \mathcal{K}_a \to \mathcal{U}$ is the 'indecomposable functor' again. The derived functors are denoted $L_i^G Q(-)$. This case has been studied in detail by Miller [**Ml2**] and Goerss [**Ge1**].

7.1.7. $\mathcal{C} = \mathcal{K}_a$, the functor from $\mathcal{K}_a$, $K \longmapsto \mathrm{Hom}_{\mathcal{K}_a}(K, \Sigma^t M^+)$ where $M \in \mathcal{U}$, $t \geq 1$ and

$$\left(\Sigma^t M^+\right)^i = \begin{cases} \mathbb{F}_p & \text{if } i = 0 \\ \left(\Sigma^t M\right)^i & \text{otherwise} \end{cases}$$

equipped with the trivial product and the obvious augmentation. One has an isomorphism

$$\mathrm{Hom}_{\mathcal{K}_a}\left(K, \Sigma^t M^+\right) \cong \mathrm{Hom}_{\mathcal{U}}\left(Q(K), \Sigma^t M\right).$$

Thus this functor takes values in the category $\mathcal{E}$. The isomorphism follows from the fact that any map (in $\mathcal{K}_a$) $K \rightarrow \Sigma^t M^+$ factors uniquely as

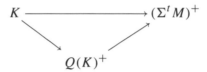

The unstable algebra $Q(K)^+$ is defined by $\left(Q(K)^+\right)^i = Q(K)^i$ if $i > 0$ and $(Q(K)^+)^0 = \mathbb{F}_p$, the product is the trivial one, and the augmentation the obvious one. Axiom $(\mathcal{K}2)$ of unstable $\mathcal{A}$-algebras holds because $\mathrm{Sq}^{|x|} x = 0$ (resp. $P^{|x|/2} x = 0$) if $p = 2$ (resp. if $p > 2$) in $Q(K)$. The derived functors are denoted $\mathrm{Ext}_{\mathcal{K}_a}(K, \Sigma^t M^+)$. Note that if K is connected it makes no difference whether we work in $\mathcal{K}$ or in $\mathcal{K}_a$.

7.2. André-Quillen cohomology of unstable $\mathcal{A}$-algebras

Let K be a connected augmented unstable $\mathcal{A}$-algebra. The derived functors $L_i^G Q(K)$ are unstable $\mathcal{A}$-modules (7.1.6). Consider now K as an object of $\mathcal{Alg}_a$. We have also derived functors $L_i^S Q(K)$ (with a small abuse of notation). We claim that, as graded vector spaces, these are isomorphic to the preceding ones. This follows from the preceding section and from the fact that those objects $G(L)$, $L \in \mathcal{K}_a$, that are connected, are always free as objects of $\mathcal{Alg}_a$. They are symmetric algebras in the graded sense. Therefore

Proposition 7.2.1 [Ml2]. *For any connected augmented unstable $\mathcal{A}$-algebra K, there is an isomorphism of graded vector spaces*

$$L_i^G Q(K) \cong L_i^S Q(K).$$

Next, if we consider $L_i^G Q(K)$ as the i-th homology group of the standard resolution of K, it supports an unstable $\mathcal{A}$-module structure. Indeed if M is an unstable $\mathcal{A}$-module, $S(M)$ is also, in a natural way, an unstable $\mathcal{A}$-module. One checks easily that the complex $\{S_{\bullet}(K), \Sigma(-1)^i d_i\}$ is a complex of unstable $\mathcal{A}$-modules. This is not a complex of unstable $\mathcal{A}$-algebras, for the product of $S(M)$ satisfies the Cartan formula although the second condition: $\mathrm{Sq}_0 x = x^2$ if $p = 2$ and $P_0 x = x^p$ if $p > 2$, does not hold.

The algebras $S_{\bullet}(K)_n$ are canonically augmented. If $\varepsilon : K \to \mathbb{F}_p$ is the augmentation of K the algebra map

$$S_{\bullet}(K)_n \xrightarrow{d_{i_n}} S_{\bullet}(K)_{n-1} \xrightarrow{d_{i_{n-1}}} \ldots S_{\bullet}(K)_0 \xrightarrow{\eta_K} K \xrightarrow{\varepsilon} \mathbb{F}_p$$

does not depend on the choice of $i_n, \ldots, i_1$ and defines a canonical augmentation on $S_{\bullet}(K)_n$. As the Cartan formula holds, the complex

$$\{Q S_{\bullet}(K), \Sigma(-1)^i Q(d_i)\}$$

is a complex of unstable $\mathcal{A}$-modules.

It is not clear, *a priori*, that the induced $\mathcal{A}$-module structure on $L_i^S Q(K)$ is the same as the natural one. However we have

Proposition 7.2.2. *The unstable $\mathcal{A}$-modules $L_i^G Q(K)$ and $L_i^S Q(K)$ are isomorphic.*

Proof. We need a more 'rigid' isomorphism. We proceed as follows [Ml2][LS1]. Consider the bisimplicial unstable $\mathcal{A}$-module $Q S_{\bullet} G_{\bullet}(K)$ defined by $Q S_{\bullet} G_{\bullet}(K)_{n,m} = Q S^{n+1}\big(G^{m+1}(K)\big)$, and form the associated double chain complex.

Filter by the degree in S. As $G_{\bullet}(K)_n$ is a symmetric algebra and, therefore, S-free, one gets in the associated spectral sequence

$$E_{s,t}^1 = \begin{cases} Q G_{\bullet}(K)_t & \text{if } s = 0, \\ 0 & \text{if } s > 0. \end{cases}$$

and

$$E_{s,t}^2 = \begin{cases} L_t^G Q(K) & \text{if } s = 0, \\ 0 & \text{if } s > 0. \end{cases}$$

Filter by the degree in G. Observe that the functor $QS_{\bullet}(K)_s$, $s \geq 1$, is equivalent to $S_{\bullet}(K)_{s-1}$. Therefore, it factors through the forgetful functor $\mathcal{K} \to \mathcal{E}_{gr}$. Using the comment after Lemma 7.1.2, one gets in the associated spectral sequence

$$E_1^{s,t} = \begin{cases} QS_{\bullet}(K)_s & \text{if } t = 0, \\ 0 & \text{if } t > 0 \end{cases}$$

and

$$E_2^{s,t} = \begin{cases} L_s^S Q(K) & \text{if } t = 0, \\ 0 & \text{if } t > 0. \end{cases}$$

As $E_{0,t}^2$ and $E_2^{0,t}$ are isomorphic, as $\mathcal{A}$-module, to the t-th homology of the total complex we are done.

Note the following important.

Corollary 7.2.3. *Let K be a connected augmented unstable $\mathcal{A}$-algebra which is locally finite as an $\mathcal{A}$-module. Then, for any i, the unstable $\mathcal{A}$-module $L_i^G Q(K)$ is also locally finite.*

We shall see in 7.6 that one can remove the connectivity hypothesis.

Proof.   The category $\mathcal{B}$ of locally finite unstable $\mathcal{A}$-modules is closed under tensor products, quotients and colimits. Therefore if M is in $\mathcal{B}$ so is $S(M)$. The result follows.

We have

Proposition 7.2.4. *Let K be an augmented unstable $\mathcal{A}$-algebra. Then the following conditions are equivalent:*

(i) $Q(K)$ *is locally finite,*

(ii) $L_i^G Q(K)$ *is locally finite for any i.*

Proof.   It is enough to prove that (i) $\Rightarrow$ (ii). Proposition 3.9.7 (i) implies that K is isomorphic to the component, $T_V(K)_t$, of the trivial map t of $T_V(K)$ for any V. Moreover let $\varepsilon : L \to \mathbb{F}_p$ be an augmented algebra, and let $L_\varepsilon = L \bigotimes_{L^0} \mathbb{F}_p$. It will be proved in (7.6.1) that, as $\mathcal{A}$-modules, $L_i^S Q(L)$ is isomorphic to $L_i^S Q(L_\varepsilon)$ for any i.

As $T_V S(K)$ is isomorphic to $S(T_V K)$, one checks easily that $T_V S_{\bullet}(K)$ is a simplicial resolution of $T_V K$. Then, using exactness of T_V, one gets the isomorphism:

$$T_V L_i^S Q(K) \cong L_i^S Q\left(T_V(K)\right).$$

Then:

$$L_i^S Q(T_V(K)) \cong L_i^S Q(T_V(K))_t \cong L_i^S Q(K).$$

And we are done by (6.2.1).

7.3. A connectivity result for André-Quillen cohomology

The main results of this section are the following two connectivity results.

Proposition 7.3.1 [L4]. *Let C be a connected $\mathbb{N}$-graded $\mathbb{F}_p$-algebra. Then the $\mathbb{N}$-graded $\mathbb{F}_p$-vector space $L_i^S Q(C)$ is i-connected (trivial in degrees less or equal to i).*

Because of the results of the preceding section this applies to $K \in \mathcal{K}$ and $L_i^G Q(K)$.

Proposition 7.3.2 [L1]. *Let K be a connected unstable $\mathcal{A}$-algebra. Then the $\mathbb{F}_p$-vector space*

$$\mathrm{Ext}_{\mathcal{K}}^s(K, \Sigma^t \mathbb{F}_p{}^+)$$

is trivial if $0 < t \leq s$.

The main ingredient of the proof of these two propositions is the following technical result on simplicial resolutions whose proof will be given in the next section.

Proposition 7.3.3 [L4]. *Let C be a connected ($C^0 = \mathbb{F}_p$) $\mathbb{N}$-graded $\mathbb{F}_p$-algebra (resp. unstable $\mathcal{A}$-algebra). Then C has in the category $\mathcal{A}lg$ (resp $\mathcal{K}$) a simplicial resolution $C_{\bullet}$ such that $C_0^0 \cong \mathbb{F}_p$ and such that the following holds: any element $x \in C_n$, $n > 0$, such that $|x| \leq n$ is degenerate.*

Recall that in an abelian simplicial group S a degenerate element $x \in S_n$ is a sum of elements of the form $s_i y$ for $y \in S_{n-1}$, $0 \le i \le n - 1$.

Before proving 7.3.1 and 7.3.2 it is worth recalling Moore's normalization theorems [M]. Let $G_\bullet$ be a simplicial abelian group (in our case it will be a simplicial $\mathbb{F}_p$-vector space). The n-th homotopy group, $\pi_n(G_\bullet)$, of $G_\bullet$ is defined to be the n-th homology group of the chain complex $\{G_n, \Sigma(-1)^i d_i\}$, $n \in \mathbb{N}$.

7.3.4. Let $N_n(G_\bullet)$ be the intersection $\bigcap_{0 \le i \le n-1} \operatorname{Ker} d_i \subset G_n$ and let $\bar{\pi}_n(G_\bullet)$ be the n-th homology group of the chain complex:

$$\to N_n(G_\bullet) \xrightarrow{d_n} N_{n-1}(G_\bullet) \to \quad .$$

The first form of the normalization theorem says that the canonical map $\bar{\pi}_n(G_\bullet) \to \pi_n(G_\bullet)$ is an isomorphism for any n.

7.3.5. Let $D(G_\bullet)_n \subset G_n$ be the subgroup generated by degenerate elements and let $\pi_n^*(G_\bullet)$ be the n-th homology group of the chain complex:

$$\to G_n/D_n(G_\bullet) \xrightarrow{\Sigma(-1)^i d_i} G_{n-1}/D_{n-1}(G_\bullet) \to \quad .$$

The second form of the normalization theorem says that the canonical map $\pi_n(G_\bullet) \to \pi_n^*(G_\bullet)$ is an isomorphism for any n.

7.3.6. In fact the result is a bit more precise (see also [Do]). The chain complex $A(G_\bullet) = \{G_n, \Sigma(1)^i d_i\}$, $n \in \mathbb{N}$, is the direct sum of the chain complexes $N(G_\bullet) = \{N_n(G_\bullet), (-1)^n d_n\}$, $n \in \mathbb{N}$, and $D(G_\bullet) = \{D_n(G_\bullet), \Sigma(-1)^i d_i\}$, $n \in \mathbb{N}$. Moreover $D(G_\bullet)$ is acyclic.

Proof of 7.3.1. Consider the simplicial $\mathbb{N}$-graded $\mathbb{F}_p$-vector space $Q(C_\bullet)$. In $Q(C_n)$, $n > 0$, all elements of internal degree less or equal to n are degenerate and $Q(C_0)^0 = 0$. The result follows from (7.3.5).

Proof of 7.3.2. The argument is the same as above. Let $K_\bullet$ be a simplicial resolution of K. We have to compute the s-th cohomology group of the cochain complex associated to the cosimplicial $\mathbb{F}_p$-vector space $\mathrm{Hom}_{\mathcal{U}}(Q(K_\bullet), \Sigma^t \mathbb{F}_p)$.

Recall that a cosimplicial object in a category $\mathcal{C}$ is a covariant functor from the category Δ (3.2) to $\mathcal{C}$. Alternatively it is a family of objects C^n, $n \geq 0$, provided with morphisms $d^j : C^{n-1} \to C^n$, $0 \leq j \leq n$ and $s^j : C^n \to C^{n-1}$, $0 \leq j \leq n-1$ satisfying the identities which are dual of those satisfied by the simplicial maps:

(7.3.7)
$$
\begin{aligned}
d^j d^i &= d^i d^{j-1} && \text{if } i < j, \\
s^j d^i &= d^i s^{j-1} && \text{if } i < j, \\
s^j d^i &= \mathrm{Id} && \text{if } i = j, j+1, \\
s^j d^i &= d^{i-1} s^j && \text{if } i > j+1, \\
s^j s^i &= s^{i-1} s^j && \text{if } i > j.
\end{aligned}
$$

In this context, the normalization theorem takes the following form. If $G^\bullet$ is a cosimplicial abelian group, its t-th cohomotopy group $\pi^t(G^\bullet)$ is the t-th cohomology group of the cochain complex $\{G^n, \Sigma(-1)^i d^i\}$, $n \in \mathbb{N}$, It is isomorphic -via the canonical map- to the t-th cohomology group of the cochain subcomplex

$$
N(G^\bullet) = \Big\{ \bigcap_{0 \leq i \leq n-1} \mathrm{Ker}\, s^i, \Sigma(-1)^i d^i \Big\}_{n \in \mathbb{N}}.
$$

In our case the map s^i is induced by composition with s_i. As $D_n Q(K_\bullet)$ is the sum of the subspaces $\mathrm{Im}\, s_i$ we get:

$$
N^n(\mathrm{Hom}_{\mathcal{U}}(Q(K_\bullet), \Sigma^t \mathbb{F}_p) = \mathrm{Hom}_{\mathcal{U}}(Q(K_n)/D_n Q(K_\bullet), \Sigma^t \mathbb{F}_p).
$$

Note that $N_n Q(K_\bullet)$ and $D_n Q(K_\bullet)$ are $\mathcal{A}$-modules. Now, as all elements in K_n of internal degree less or equal to n are degenerate the same is true for $Q(K_n)$. Therefore, $Q(K_n)/D_n Q(K_\bullet) \cong N_n Q(K_\bullet)$ has connectivity n and $\mathrm{Ext}^s_{\mathcal{K}}(K, \Sigma^t \mathbb{F}_p{}^+)$ is trivial if $s \geq t$.

7.4. Proof of Proposition 7.3.3

We start with the case of $C \in \mathcal{A}lg$. The following proof is an adaptation of André's 'step by step' method. It is just the classical method of 'adding' cells in order to kill homotopy groups. We have to construct a simplicial resolution $C_\bullet$ of C such that $N_n(C_\bullet)$ has connectivity n for all $n > 0$ and such that $C_0^0 \cong \mathbb{F}_p$.

One needs the following:

Proposition 7.4.1. *Let $C_\bullet$ be a simplicial $\mathbb{N}$-graded $\mathbb{F}_p$-algebra. Assume that $C_n^0 \cong \mathbb{F}_p$ for any n and that each C_n is provided with the obvious augmentation. Assume moreover that the $\mathbb{N}$-graded $\mathbb{F}_p$-vector space $N_n(Q(C_\bullet))$ has connectivity n ($N_n(Q(C_\bullet))^i = 0$ if $i \leq n$ for any n). Then the $\mathbb{N}$-graded $\mathbb{F}_p$-vector space $N_n(C_\bullet)$ has connectivity n for all $n > 0$. Moreover the canonical map $N_n(C_\bullet) \to N_n(Q(C_\bullet))$ is an isomorphism in degree $n + 1$.*

For $n = 0$, the unit of C_0 is not degenerate. However, $Q(C_0)^0 = 0$ for C_0 is connected.

In view of Propositions 7.4.1 and 7.3.6 in order to prove Proposition 7.3.3 it is enough to construct a simplicial resolution $C_\bullet$ of C such that $N_n Q(C_\bullet)$ has connectivity n for all n. We prove (7.4.1) and shall come back after to (7.3.3).

Proof of 7.4.1. Consider a finite family of elements $x_i \in C_{\alpha_i}$, we assume that $|x_i| > \alpha_i$. Consider also a family of monotonic maps $\mu_i : [m] \to [\alpha_i]$, where m is a given integer.

We claim that the element

$$\Pi_i \mu_i^* x_i \in C_m^{\Sigma |x_i|}$$

is degenerate as soon as either $m \geq \sum_i |x_i|$ or $m + 1 \geq \sum_i |x_i|$ and there are at least two terms in the sum.

We shall prove this claim below. Our assumptions on $C_\bullet$ imply that any element $x \in C_n$ such that $|x| \leq n$, or such that $|x| = n + 1$ and x is decomposable is a sum of elements as just described. This follows from a double induction over n and $|x|$. Therefore the claim implies (7.4.1).

The claim will follow from:

Lemma 7.4.2. *Let m be a given integer, let $\{\mu_i : [m] \to [\alpha_i]\}_{i \in I}$ be a finite family of monotonic maps. Assume that $m > \sum_i \alpha_i$. Then there exists $k \in [m]$ such that $\mu_i(k) = \mu_i(k + 1)$ for all i. Thus all maps μ_i^* factors as $s_k \circ \widetilde{\mu}_i^*$.*

Proof of 7.4.2. It is enough to show that for some $0 \leq \ell \leq m$ the sum $\sum_i (\mu_i(\ell+1) - (\mu_i(\ell))$ is zero. But

$$\sum_{0 \leq \ell \leq m-1} \sum_i (\mu_i(\ell+1) - \mu_i(\ell)) = \sum_i \mu_i(m) - \mu_i(0)$$

$$\leq \sum_i \alpha_i \, .$$

As m is strictly greater than $\sum \alpha_i$ this implies the lemma.

Let us come back to the claim. If $m \geq \sum_i |x_i|$ the lemma directly implies that the element $\Pi_i \mu_i^* x_i \in C_m^{\Sigma |x_i|}$ is degenerate. If $m + 1 = \sum_i |x_i|$, as this sum sum involves at least two (non-zero) terms, one has:

$$\sum_{0 \leq \ell \leq m-1} \sum_i (\mu_i(\ell+1) - \mu_i(\ell)) \leq \alpha_1 + \alpha_2 + \ldots \leq -2 + \sum_i |x_i| \leq m - 1 \, .$$

And the result follows as above.

We now describe the step by step method of André. This is a way to construct inductively a simplicial resolution having the property we are looking for. We shall construct it in (7.4.6). Our starting point is a simplicial $\mathbb{N}$-graded $\mathbb{F}_p$-algebra $C_\bullet$ such that for a given integer $n \geq 1$ the following set of conditions, denoted by H_n, holds:

7.4.3.

(i) $C_m^0 \cong \mathbb{F}_p$ for any m;

(ii) $N_m(C_\bullet)$ is m-connected for all m;

(iii) $\pi_m(C_\bullet)$ is trivial for $0 < m < n$.

Note that the normalization theorem implies that $\pi_m(C_\bullet)$ is m-connected for all m. We are going to construct a simplicial $\mathbb{N}$-graded $\mathbb{F}_p$-algebra $\bar{C}_\bullet$ with a map $c : C_\bullet \to \bar{C}_\bullet$ such that conditions H_{n+1} hold and moreover such that:

7.4.4.

(i) $c_m : C_m \to \bar{C}_m$ is injective for any m, bijective if $m \leq n$;

(ii) $\bar{C}_m$ is a symmetric algebra over C_m for all m.

7.4.5. Construction of $\bar{C}_\bullet$.   Let $\Delta[n]$ be the standard n-simplex and let i_n be the unique non degenerate n-simplex of $\Delta[n]$. Let $\Delta[n, k]$ be the simplicial subset generated by the simplices $d_0 i_n, \ldots,$ $d_{k-1} i_n, d_{k+1} i_n, \ldots, d_n i_n$.

Denote by $C[n, k]$ the quotient $\Delta[n]/\Delta[n, k]$. The image of $\Delta[n, k]$ in $C[n, k]$ is chosen as base point. Let $\mathbb{F}_p C[n, k]_*$ be the free simplicial $\mathbb{F}_p$-vector space freely generated by $C[n, k]$ modulo the relation base point $= 0$. The set of homomorphisms from $\mathbb{F}_p C[n, n]_*$ to a simplicial $\mathbb{F}_p$-vector space $E_\bullet$ is, by definition, isomorphic to $N_n E_\bullet$.

Let $\Delta[n]/\overset{\circ}{\Delta}[n]$ be the 'simplicial n-sphere': the quotient of $\Delta[n]$ by the simplicial subset generated by $d_0 i_n, \ldots, d_n i_n$. And let $\mathbb{F}_p \Delta[n]/\overset{\circ}{\Delta}[n]_*$ be defined as above.

Consider the simplicial $\mathbb{N}$-graded $\mathbb{F}_p$-vector space

$$(\mathbb{F}_p \Delta[n]/\overset{\circ}{\Delta}[n]_*) \otimes \pi_n C_\bullet ,$$

the simplicial $\mathbb{N}$-graded $\mathbb{F}_p$-vector space

$$((\mathbb{F}_p C[n + 1, n + 1]_*) \otimes \pi_n C_\bullet) \oplus C_\bullet ,$$

and the map γ from the first one to the second one which is the direct sum of the two following maps
— the map

$$\mathbb{F}_p \Delta[n]/\overset{\circ}{\Delta}[n]_* \otimes \pi_n C_\bullet \to \mathbb{F}_p C[n + 1, n + 1]_* \otimes \pi_n C_\bullet$$

which is determined by $i_n \otimes y \mapsto -d_{n+1} i_{n+1} \otimes y$;
— the map

$$\mathbb{F}_p \Delta[n]/\overset{\circ}{\Delta}[n]_* \otimes \pi_n(C_\bullet) \to C_\bullet$$

defined by $i_n \otimes y \mapsto s_n(y)$, where s_n is a section of the projection from $\mathrm{Ker}\, d_n \subset N_n C_\bullet$ onto $\pi_n(C_\bullet)$.

Let $C'_\bullet$ be the cokernel of γ. We define $\bar{C}_\bullet$ as follows. The algebra $\bar{C}_m$ is the quotient of the symmetric algebra over C'_m modulo the relations $x \otimes y = xy$ whenever x and y belong to the image of C_m in C'_m (xy denotes the product in C_m). This is the symmetric algebra, over C_m, generated by C'_m/C_m (as graded $\mathbb{F}_p$-vector space): $C_m \otimes S(C'_m/C_m)$.

It is clear that the map $C_m \to C'_m$ is injective for any m and bijective for $m \leq n$. This is left to the reader. Therefore 7.4.4 (i) and (ii) hold for $\bar{C}_{\bullet}$.

It remains to prove that conditions H_{n+1} (see 7.4.3) hold. Here (i) is obvious, (ii) follows from (7.4.1) and H_n, and (iii) holds by construction: the map $d_{n+1} : N_{n+1}(C'_{\bullet}) \to N_n(\bar{C}_{\bullet}) = N_n(C_{\bullet})$ is surjective on $s_n(\pi_n(C_{\bullet}))$.

7.4.6. End of proof of 7.3.3. The proof is done by induction. Consider a symmetric algebra C_0 that surjects onto C. We can assume that $C_0 \to C$ is an isomorphism in degree 1. Therefore the kernel I of $C_0 \to C$ is 1-connected. Let $C_{0,\bullet}$ be the constant simplicial algebra equal to C_0 in each degree. The conditions H_0 hold for $C_{0,\bullet}$. One observes that H_1 holds for $C_{1,\bullet} = \bar{C}_{0,\bullet} \cdots$

One can define $C_{2,\bullet} = \bar{C}_{1,\bullet}, \ldots C_{n,\bullet} = \bar{C}_{n-1,\bullet}$. One defines $C_{\bullet}$ by $C_m = C_{n,m}$ for $n \geq m$.

By construction:

— $C_{\bullet}$ is augmented over C;

— C_n is a symmetric algebra; the complex of graded $\mathbb{F}_p$-vector spaces

$$\to C_n \xrightarrow{\Sigma(-1)^i d_i} C_{n-1} \to \ldots \to C_0 \to C \to 0$$

is acyclic. The graded $\mathbb{F}_p$-vector space $N_n(Q(C_{\bullet}))$ is n-connected. Therefore we have built up a resolution satisfying the assumptions of Proposition 7.4.1. We are done.

7.4.7 The case of the category $\mathcal{K}$. The above proof extends without difficulties to $\mathcal{K}$. It is enough to replace in the construction from $C'_{\bullet}$ to $C_{\bullet}$ the symmetric algebra functor by the free unstable $\mathcal{A}$-algebra. All the rest goes through the same way.

7.5. Miller's spectral sequence

In [Ml2], Miller introduces a spectral sequence which converges to $\mathrm{Ext}^s_{\mathcal{K}_a}(K, \Sigma M^+)$ (ΣM^+ is interpreted as in 7.1.7). This spectral sequence allows to study separately the effects on these groups of the algebra product of K and of the $\mathcal{A}$-module structure of K. This spectral sequence allows us to recover (7.3.2) from (7.3.1). Although it is not essential in the presentation chosen here, it is interesting by itself, and we reproduce it.

Theorem [Ml2]. *There is a cohomological spectral sequence with* $E_2^{p,q}$-*term isomorphic to* $\mathrm{Ext}^p_{\mathcal{U}}(L^G_q \Omega Q(K), M)$ *converging to* $\mathrm{Ext}^{p+q}_{\mathcal{K}_a}(K, \Sigma M^+)$.

Proof. We have observed that $\mathrm{Hom}_{\mathcal{K}_a}(K, \Sigma M^+)$ is isomorphic to $\mathrm{Hom}_{\mathcal{U}}(Q(K), \Sigma M)$, hence to $\mathrm{Hom}_{\mathcal{U}}(\Omega Q(K), M)$. Therefore, the functor $K \mapsto \mathrm{Hom}_{\mathcal{K}_a}(K, \Sigma M^+)$ factors as $\mathrm{Hom}_{\mathcal{U}}(-, M) \circ \Omega Q$.

Let $K_\bullet$ be a resolution (in $\mathcal{K}$) of K. Consider the chain complex:

$$\to \Omega Q(K_n) \xrightarrow{\Sigma(-1)^i d_i} \Omega Q(K_{n-1}) \to$$

Choose a projective resolution of this complex of unstable $\mathcal{A}$-modules [Gd]. We have complexes of unstable $\mathcal{A}$-modules $L^i_\bullet$, $i \geq 0$ with maps of complexes of $\mathcal{A}$-modules $f_i : L^i_\bullet \to L^{i-1}_\bullet$ and a map (of complexes of $\mathcal{A}$-modules) $f_0 : L^0_\bullet \to \Omega Q(K_\bullet)$ such that:

— each L^i_k is projective as an unstable $\mathcal{A}$-module;

— $Z_k(L^i_\bullet)$, $B_k(L^i_\bullet)$ and $H_k(L^i_\bullet)$ are projective unstable $\mathcal{A}$-modules, for all i and $k \geq 0$;

— the complexes:

$$\longrightarrow L^i_k \longrightarrow L^{i-1}_k \longrightarrow \ldots \longrightarrow \Omega Q(K_k) \longrightarrow 0,$$

$$\longrightarrow Z_k L^i_\bullet \longrightarrow Z_k L^{i-1}_\bullet \longrightarrow \ldots \longrightarrow Z_k \Omega Q(K_\bullet) \longrightarrow 0,$$

$$\longrightarrow B_k L^i_\bullet \longrightarrow B_k L^{i-1}_\bullet \longrightarrow \ldots \longrightarrow B_k \Omega Q(K_\bullet) \longrightarrow 0,$$

$$\longrightarrow H_k L^i_\bullet \longrightarrow H_k L^{i-1}_\bullet \longrightarrow \ldots \longrightarrow H_k \Omega Q(K_\bullet) \longrightarrow 0,$$

are projective resolutions of unstable $\mathcal{A}$-modules.

Consider the double cochain complex $\left\{\mathrm{Hom}_{\mathcal{U}}(L_p^q, M)\right\}_{p,q \geq 0}$.
Filter with respect to p:

$$E_1^{p,q} = \begin{cases} \mathrm{Hom}_{\mathcal{U}}(\Omega Q(K_p), M) & \text{if } q = 0, \\ 0 & \text{if } q > 0. \end{cases}$$

This is because $\Omega Q(K_p)$ is projective in $\mathcal{U}$ (recall that K is free in $\mathcal{K}$). Therefore

$$E_2^{p,q} = \begin{cases} \mathrm{Ext}_{\mathcal{K}_a}^p(K, \Sigma M^+) & \text{if } q = 0, \\ 0 & \text{if } q > 0. \end{cases}$$

Filter with respect to q: $E_1^{p,q} \cong \mathrm{Hom}_{\mathcal{U}}(Z_p L^q / B_p L^q, M)$. Therefore, $E_2^{p,q}$ is isomorphic to $\mathrm{Ext}_{\mathcal{U}}^q(L_p^G \Omega Q(K), M)$. The result follows.

7.6. A change of rings theorem

Let K be an object of $\mathcal{K}_a$, $\varepsilon : K \to \mathbb{F}_p$ be the augmentation. Let K_ε be $K \otimes_{K^0} \mathbb{F}_p$. The flat base change theorem of André and Quillen [An][Q1] is:

Theorem 7.6.1. *There is an isomorphism of unstable $\mathcal{A}$-modules*

$$L_i^G Q(K) \cong L_i^G Q(K_\varepsilon) \quad \text{for all } i \geq 0.$$

Moreover [L1]:

Theorem 7.6.2. *For any unstable $\mathcal{A}$-module M there is an isomorphism of $\mathbb{F}_p$-vector spaces*

$$\mathrm{Ext}_{\mathcal{K}_a}^s(K, \Sigma M^+) \cong \mathrm{Ext}_{\mathcal{K}_a}^s(K_\varepsilon, \Sigma M^+),$$

for all $s \geq 0$.

We state a theorem only in the context of the category $\mathcal{K}_a$ because we shall use that, for any unstable $\mathcal{A}$-algebra K, K^0 is a p-Boolean algebra. A p-Boolean algebra is a free unstable $\mathcal{A}$-algebra or a filtered colimit of free unstable $\mathcal{A}$-algebras, *i.e.* a 'flat' unstable $\mathcal{A}$-algebra, and this is crucial in the proof of 7.6.1.

It is enough to prove 7.6.1 and 7.6.2 in the case where K^0 is a finite dimensional $\mathbb{F}_p$-vector space. Because if K is any object of $\mathcal{K}_a$, the standard resolution of K is the colimit of the standard resolutions of its unstable subalgebras K_α such that $\dim_{\mathbb{F}_p} K_\alpha^0 < +\infty$.

Let $\varepsilon_0, \ldots, \varepsilon_n$ be all the augmentation homomorphisms of K (all graded ring homorphisms from K, or K^0, to $\mathbb{F}_p$). Suppose that the given augmentation is ε_0. Let K_i be the unstable $\mathcal{A}$-algebra $K \otimes_{K^0} \mathbb{F}_p$, where $\mathbb{F}_p$ is provided with the K^0-module structure given by ε_i. The unstable $\mathcal{A}$-algebra K is isomorphic to the product $\prod_i K_i$. Let $K_{i,\bullet}$ be a $\mathcal{K}$-free resolution of K_i. We can assume that the two following conditions hold:

— $K_{i,n} \cong K_{j,n}$ for all i, j and n, but the faces and degeneracies depend on i and j,

— $(K_{i,n})^0 \cong \mathbb{F}_p$ for any i and n.

To prove it, observe that, in the step by step construction, one first chooses a free unstable $\mathcal{A}$-algebra that surjects onto K. However, if we have a finite family K_i it is possible to choose a free unstable $\mathcal{A}$-algebra that surjects onto each K_i. Constructing in this fashion each resolution $K_{i,\bullet}$ simultaneously one observes that at each step it is enough 'to choose' a 'sufficiently large free unstable $\mathcal{A}$-algebra'.

Denote by L_n the common value of the $K_{i,n}$. Clearly $\prod_i K_{i,n}$ is isomorphic to $K^0 \otimes L_n$ as an unstable $\mathcal{A}$-algebra, and is free (observe that a free unstable $\mathcal{A}$-algebra of finite type L is isomorphic to $L^0 \otimes \bar{L}$ for a certain connected free unstable $\mathcal{A}$-algebra $\bar{L}$).

Therefore, $\prod_i K_{i,\bullet}$ is a free $\mathcal{K}$-resolution of K. It becomes a free $\mathcal{K}_a$-resolution by defining an augmentation as follows. One requires the augmentation to be trivial on $K_{i,\bullet}$ if $i \geq 1$ and to extend ε on $K_{0,\bullet}$. Recall that as $\varepsilon_0 = \varepsilon$ $K_\varepsilon = K_0$. Then $Q(\prod_i K_{i,\bullet}) \cong Q(K_{0,\bullet})$ and

$$\mathrm{Hom}_{\mathcal{K}_a}\left(\prod_i K_{i,\bullet}, \, \Sigma M^+\right) = \mathrm{Hom}_{\mathcal{K}_a}\left(K_{0,\bullet}, \, \Sigma M^+\right),$$

and we are done.

In fact Theorem 7.6.2 can also be deduced from Theorem 7.6.1 and Miller's spectral sequence.

Corollary 7.6.3. *For any augmented unstable $\mathcal{A}$-algebra K the unstable $\mathcal{A}$-module $L_i^G Q(K)$ is in $\mathcal{N}il_i$.*

We can restrict attention to the case when $K_0 = \mathbb{F}_p$ by Theorem 7.6.1. Let $K_\bullet$ be a simplicial resolution of K. Then $T_V K_\bullet$ is a simplicial resolution of $T_V K$. In order to prove this one uses the exactness of T_V and the formula $T_V(U(M)) \cong U(T_V(M))$. It follows by Theorem 7.6.1 that $T_V L_i^G Q(K)$ is isomorphic $L_i^G(Q(T_V(K)_{T_V(\varepsilon)}))$, Section 7.3 shows that $L_i^G Q(K)$ and $L_i^G Q(T_V(K)_{T_V(\varepsilon)})$ are i-connected. The result follows from (6.1.1).

We note also the following obvious corollary of Theorem 7.6.2 and Corollary 7.2.3

Corollary 7.6.4. *Let K be an augmented unstable $\mathcal{A}$-algebra which is locally finite as an $\mathcal{A}$-module. Then, for any i, the unstable $\mathcal{A}$-module $L_i^G Q(K)$ is also locally finite.*

7.7. Derivations

In section 7.8 we shall prove a theorem of Lannes which says that certain groups that occur as the E_2-term of a certain Bousfield-Kan spectral sequence are trivial. These groups are going to be defined in this section. The cancellation theorem will be used in the next chapter to prove Miller's conjecture.

Let $\mathcal{K}/K$ be the category of unstable $\mathcal{A}$-algebra over a given unstable $\mathcal{A}$-algebra K. The objects are maps $\varphi : L \to K$ in $\mathcal{K}$; a morphism from $\varphi_0 : L_0 \to K$ to $\varphi_1 : L_1 \to K$ is a map $\psi : L_0 \to L_1$ in $\mathcal{K}$ such that $\varphi_0 = \varphi_1 \circ \psi$.

Let M be a K-module, which is also an unstable $\mathcal{A}$-module and such that the structure map $K \otimes M \to M$ is $\mathcal{A}$-linear. We let $\mathcal{U}_K$ denote the category of such objects. Let $\varphi : L \to K$ be in $\mathcal{K}/K$ and let M be in $\mathcal{U}_K$. Define a product on $K \oplus M$ by the formula:

$$(k + m)(k' + m') = kk' + (km' + (-1)^{|k'||m|}k'm),$$

for any $k, k' \in K$ and $m, m' \in M$. This defines an unstable $\mathcal{A}$-algebra structure on $K \oplus M$ if and only if the following holds:

$$Sq^{|m|}m = 0 \quad \text{for all} \quad m \in M \quad \text{if} \quad p = 2,$$
$$P^{|m|/2}m = 0 \quad \text{for all} \quad m \in M \quad \text{of even degree if} \quad p > 2.$$

If p is equal to 2 this means that M is a suspension; if $p > 2$ the condition is weaker. We shall denote by $\mathcal{V}_K$ the full subcategory of $\mathcal{U}_K$ whose objects satisfy this condition ($\mathcal{V}$ if $K = \mathbb{F}_p$).

Then (compare with [Q1])

Proposition 7.7.1. *Let M be in $\mathcal{V}_K$. Then $K \oplus M$, provided with the projection to K, is an abelian object of $\mathcal{K}/K$, i.e.*

$$\mathrm{Hom}_{\mathcal{K}/K}(L, K \oplus M)$$

is an abelian group for any $L \in \mathcal{K}/K$, and in fact an $\mathbb{F}_p$-vector space.

Conversely any abelian group object of $\mathcal{K}/K$ is of this form.

We keep the preceding notations.

Definition 7.7.2. The $\mathbb{F}_p$-vector space of ($\mathcal{A}$-linear) K-derivations from L into M is the $\mathbb{F}_p$-vector space of maps $\mathrm{Hom}_{\mathcal{K}/K}(L, K \oplus M)$. Alternatively, we shall denote it $\mathrm{Der}_{\mathcal{K}/K}(L, M)$.

In other words, we are looking at $\mathcal{A}$-linear maps $u : L \to M$ such that $u(xy) = \varphi(x)u(y) + (-1)^{|x||y|}\varphi(y)u(x)$ for any x and y in L. Note that in the preceding notation, the map $\varphi : L \to K$ is implicit.

We now explain how derivations occur in the context of the cohomology of mapping spaces. Consider the mapping space $\mathrm{map}(X, Y)$. In studying this space with the Bousfield-Kan spectral sequence (to be described later) derivations occur as follows. Let $f \in \pi_0\mathrm{map}(X, Y)$ and u be in $\pi_t\mathrm{map}(X, Y)_f$, $t > 0$, where $\mathrm{map}(X, Y)_f$ denotes the connected component of f in $\mathrm{map}(X, Y)$. We have a diagram

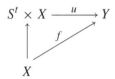

which commutes up to homotopy. We get maps

$$u^* : H^*Y \to H^*S^t \otimes H^*X,$$
$$f^* : H^*Y \to H^*X.$$

Therefore, we can consider H^*Y as an object of $\mathcal{K}/H^*X$, via f^*. Write $H^*S^t \otimes H^*X$ as $H^*X \oplus \Sigma^t H^*X$. Then u^* is an element of

$\mathrm{Hom}_{\mathcal{K}/H^*X}(H^*Y, H^*X \oplus \Sigma^t H^*X) = \mathrm{Der}_{\mathcal{K}/H^*X}(H^*Y, \Sigma^t H^*X)$.
Note that the H^*X-module structure on $\Sigma^t H^*X$ is given by the formula $x(\Sigma^t y) = (-1)^{t|x|}\Sigma^t xy$.

7.8. Derived functors of derivations and Lannes' theorem

The main result of this section is Theorem 7.8.3. In order to state it we need to introduce derived functors of derivations. Recall that K is an unstable $\mathcal{A}$-algebra, and that M is an object of $\mathcal{V}_K$. We want to define derived functors of

$$K \mapsto \mathrm{Der}_{\mathcal{K}/K}(K, M), \quad \mathcal{K}/K \to \mathcal{E}.$$

First, we need to define free objects in $\mathcal{K}/K$. This can be done as in 7.2. Consider the forgetful functor from $\mathcal{K}/K$ to the category $\mathcal{E}_{gr}/\mathcal{O}K$ of $\mathbb{N}$-graded $\mathbb{F}_p$-vector spaces over $\mathcal{O}K$. It has a left adjoint $\mathcal{G}_K$ and one has a comonad with functor $G_K = \mathcal{G}_K \circ \mathcal{O}$. The free objects are just free objects in $\mathcal{K}$ provided with a map to K. Now let $L_\bullet$ be a free resolution of $L \in \mathcal{K}$, and suppose moreover that we have a map $\varphi : L \to K$ (in $\mathcal{K}$). Then the composite maps

$$L_n \xrightarrow{d_{i_n}} L_{n-1} \xrightarrow{d_{i_{n-1}}} \ldots \to L_0 \xrightarrow{\varepsilon} L \xrightarrow{\varphi} K$$

do not depend on $i_1, i_2, \ldots, i_n$. Therefore $L_\bullet$ is a simplicial object over K and it is easily checked that it is a simplicial resolution in $\mathcal{K}/K$ of $\varphi : L \to K$.

Definition 7.8.1. The s-th derived functor of

$$\mathrm{Der}_{\mathcal{K}/K}(-, M) : \mathcal{K}/K \to \mathcal{E}$$

is defined to be, at $L \in \mathcal{K}/K$, the s-th cohomology group of the cochain complex

$$\left\{ \mathrm{Der}_{\mathcal{K}/K}(L_n, M); \Sigma(-1)^i d^i \right\}_{n \in \mathbb{N}},$$

where $L_\bullet$ is any free resolution of L. It is denoted by $\mathrm{Der}^s_{\mathcal{K}/K}(L, M)$.

The following crucial adjointness theorem is due to Lannes:

Theorem 7.8.2. *Let* L *be an object of* $\mathcal{K}/H^*V$. *Then, for all* $t > 0$, *there is an isomorphism:*

$$\mathrm{Der}^s_{\mathcal{K}/H^*V}(L, \Sigma^t H^*V) \cong \mathrm{Ext}^s_{\mathcal{K}_a}(T_V L, \Sigma^t \mathbb{F}_p{}^+).$$

Proof. Let $L_\bullet$ be a free resolution of L (in $\mathcal{K}$ and therefore in $\mathcal{K}/H^*V$). Then we claim that $T_V L_\bullet$ is a free resolution of $T_V L$. This follows from the following facts:

(i) if L_n is free, then so is $T_V(L_n)$; one uses (3.9.2) or the very definition of the division in the category $\mathcal{K}$;

(ii) T_V restricts to a functor from $\mathcal{K}$ into itself;

(iii) T_V is exact.

Now consider the $\mathbb{F}_p$-vector space

$$\mathrm{Der}_{\mathcal{K}/H^*V}(L, \Sigma^t H^*V) = \mathrm{Hom}_{\mathcal{K}/H^*V}(L, H^*S^t \otimes H^*V),$$

L being any object in $\mathcal{K}/H^*V$. It is easily checked to be the inverse image of the structure map $L \to H^*V$ under the map:

$$\mathrm{Hom}_{\mathcal{K}}(L, H^*S^t \otimes H^*V) \to \mathrm{Hom}_{\mathcal{K}}(L, H^*V)$$

induced by the augmentation $H^*S^t \to \mathbb{F}_p$. Thus, by definition of T_V, it is isomorphic to the inverse image of the map adjoint to $L \to H^*V$ under the map:

$$\mathrm{Hom}_{\mathcal{K}}(T_V L, H^*S^t) \to \mathrm{Hom}_{\mathcal{K}}(T_V L, \mathbb{F}_p).$$

This is by definition $\mathrm{Der}_{\mathcal{K}_a}(T_V L, \Sigma^t \mathbb{F}_p)$; here $\mathcal{K}_a = \mathcal{K}/\mathbb{F}_p$ and the augmentation $T_V L \to \mathbb{F}_p$ is the adjoint of the structure map $L \to H^*V$. It follows that the cochain complex

$$\left\{ \mathrm{Der}_{\mathcal{K}/H^*V}(L_n, \Sigma^t H^*V); \Sigma(-1)^i d^i \right\}_{n \in \mathbb{N}}$$

is isomorphic to the complex

$$\left\{ \mathrm{Der}_{\mathcal{K}_a}(T_V L_n, \Sigma^t \mathbb{F}_p); \Sigma(-1)^i d^i \right\}_{n \in \mathbb{N}}.$$

Therefore, using that $T_V L_\bullet$ is a free resolution of $T_V L$, one has

$$\mathrm{Der}^s_{\mathcal{K}/H^*V}(L, \Sigma^t H^*V) \cong \mathrm{Ext}^s_{\mathcal{K}_a}(T_V L, \Sigma^t \mathbb{F}_p{}^+).$$

The following cancellation result of Lannes **[L1]** will be used in an essential way in the next chapter.

Theorem 7.8.3. *The group* $\mathrm{Der}^s_{\mathcal{K}/H^*V}(L, \Sigma^t H^*V)$ *is trivial for any* $L \in \mathcal{K}/H^*V$ *as soon as* $0 < t \le s$ *(note that the statement 'holds' for* $t = s = 0$*, for* $\mathrm{Der}^0_{\mathcal{K}/H^*V}(L, H^*V)$ *is the one point set).*

We first use the isomorphism

$$\mathrm{Der}^s_{\mathcal{K}/H^*V}(L, \Sigma^t H^*V) \cong \mathrm{Ext}^s_{\mathcal{K}_a}(T_V L, \Sigma^t \mathbb{F}_p{}^+).$$

Denoting the augmentation $T_V L \to \mathbb{F}_p$ by ε, we have, by Theorem 7.6.2,

$$\mathrm{Ext}^s_{\mathcal{K}_a}(T_V L, \Sigma^t \mathbb{F}_p{}^+) \cong \mathrm{Ext}^s_{\mathcal{K}_a}((T_V L)_\varepsilon, \Sigma^t \mathbb{F}_p{}^+);$$

as $(T_V L)_\varepsilon$ is connected. By Proposition 7.3.2, these groups are known to be trivial if $0 < t \le s$. We are done.

8. On homotopy classes of maps from BV

8.1. Miller's conjecture

Theorem 7.8.3 together with Bousfield-Kan's and Bousfield's cosimplical techniques yields:

Theorem 8.1.1 [L1][L4]. *Let Y be a connected space. Suppose that Y is nilpotent, that $\pi_1 Y$ is finite and that H^*Y is of finite type. Then the natural map $f \mapsto f^*$, $[BV, Y] \to \mathrm{Hom}_{\mathcal{K}}(H^*Y, H^*V)$ is a bijection.*

This was conjectured by Miller in **[Ml1]**, where he proved the result when the cohomology of Y is of the form $U(M)$ for some unstable $\mathcal{A}$-module M. Lannes and Zarati proved it when Y is an infinite loopspace. We shall not fully develop the relevant cosimplicial technology here. We shall state Bousfield-Kan's and Bousfield's theorems and show how to apply them. We refer to **[BK2]** and **[Bo4]** for proofs and more general statements. In all the sequel 'space' will mean simplicial set. The category of simplicial sets will be denoted $s\mathcal{S}$.

8.2. Bousfield-Kan's and Bousfield's theorems

We need two statements about cosimplicial spaces. Recall that a cosimplicial space is a cosimplicial object in the category of simplicial sets.

Let $X^\bullet$ and $E^\bullet$ be cosimplicial spaces. One says that $E^\bullet$ is $\mathbb{F}_p$-vector space-like if each E^n is a simplicial $\mathbb{F}_p$-vector space and if the maps d^i, $i > 0$, and s^j, $j \geq 0$, are linear (note that d^0 is not supposed to be linear).

Assume that each X^n is a free transitive space under E^n. More precisely we assume that for each n there is a map of simplicial sets $E^n \times X^n \to X^n$ which restricts in each simplicial degree q to a free transitive action $E_q^n \times X_q^n \to X_q^n$. We shall say that $X^\bullet$ is affine-like if the maps d^i, $i > 0$ (again d^0 is not included), and s^j, $j \geq 0$, are affine in each simplicial degree q (i.e. the obvious diagrams commute). If $X^\bullet$ is pointed, by considering the base point of X^n as the zero vector, one can consider it as an $\mathbb{F}_p$-vector space-like cosimplicial space. It follows that $X^\bullet$ is fibrant [BK2]. In general the cosimplicial space $X^\bullet$ will not be pointed.

The hypothesis that X^n is a simplicial $\mathbb{F}_p$-affine set implies that $\pi_i(X^n; x_0)$ is an abelian group for any $i \geq 0$ which does not depend on the choice of the base point x_0. More precisely let E^n be the simplicial $\mathbb{F}_p$-vector space acting on X^n. It is classical that the set of free homotopy classes $[S^i, X^n]$ is in natural bijection with $\pi_i(E^n; 0)$ (see for example [L4]). Thus it has a natural group structure and the forgetful map $\pi_i(X^n; x_0) \to [S^i, X]$ is an isomorphism.

It follows that $\pi_t X^\bullet$ is a well defined cosimplicial $\mathbb{F}_p$-vector space and that one can define $\pi^s \pi_t X^\bullet$. Recall that if $G^\bullet$ is a cosimplicial abelian group then $\pi^s G^\bullet$ is the s-th cohomology of the cochain complex $\{G^n, \Sigma(-1)^i d^i\}$, $n \in \mathbb{N}$. Recall the standard cosimplicial simplex $\Delta^\bullet$ defined by:

(i) $\Delta^n = \Delta[n]$;

(ii) $d^j : \Delta[n] \to \Delta[n+1]$ sends i_n to $d_j i_{n+1}$;

(iii) $s^j : \Delta[n+1] \to \Delta[n]$ sends i_{n+1} to $s_j i_n$.

The total space of a cosimplicial space $X^\bullet$ is defined to be the simplicial set $\mathrm{map}(\Delta^\bullet, X^\bullet)$, where $\mathrm{map}(\Delta^\bullet, X^\bullet)_n$ is the set of maps of cosimplicial spaces from $\Delta^\bullet \times \Delta[n]$ to $X^\bullet$, where $\Delta[n]$ is understood to be a constant cosimplicial space. This total space is denoted by $\mathrm{Tot}\, X^\bullet$.

Denote by $Sk^s \Delta^\bullet$ the s-th skeleton of $\Delta^\bullet$. In codegree n it is isomorphic to the s-th skeleton of $\Delta[n]$ [BK2], [M]. The space $\mathrm{Tot}_s X^\bullet$ is defined to be $\mathrm{map}(Sk^s \Delta^\bullet, X^\bullet)$. In this formula 'map' has the same meaning as above. The inclusion $Sk^{s-1} \Delta^\bullet \subset Sk^s \Delta^\bullet$

induces a map $\mathrm{Tot}_s X^\bullet \to \mathrm{Tot}_{s-1} X^\bullet$. It is clear that $\mathrm{Tot}_0 X^\bullet \cong X^0$ and that $\mathrm{Tot}\, X^\bullet$ is isomorphic to $\lim_s \mathrm{Tot}_s X^\bullet$.

Let X be a space and $W^\bullet$ be a cosimplicial space. One has the following natural isomorphism

(8.2.1) $\mathrm{map}(X, \mathrm{Tot}\, W^\bullet) \cong \mathrm{Tot}(\mathrm{map}(X, W^\bullet))$.

On the left, 'map' stands for the internal hom in the category of spaces [**BK2**]. It is defined by

$$\mathrm{map}(X, Y)_n = \hom_{s\mathcal{S}}(X \times \Delta[n], Y)\,.$$

On the right, it denotes the cosimplicial space which, in codegree n, is equal to the internal hom in the category $s\mathcal{S}$ $\mathrm{map}(X, W^n)$. More generally one has:

(8.2.1*) $\mathrm{map}(X, \mathrm{Tot}_s W^\bullet) \cong \mathrm{Tot}_s(\mathrm{map}(X, W^\bullet))$.

Let $X^\bullet$ be a cosimplicial space. Let α be an element of $\pi^0 \pi_0 X^\bullet$ and let $X_\alpha^\bullet$ be the cosimplicial subspace of $X^\bullet$ consisting in codegree n of the connected components of any iterated coface of some representative of α. One has $\bigsqcup_\alpha \mathrm{Tot} X_\alpha^\bullet = \mathrm{Tot} X^\bullet$ ([**Bo4**] see also [**L4**]). Moreover if $X^\bullet$ is fibrant so is $X_\alpha^\bullet$ for any α.

If $X^\bullet$ is pointed, its base point yields a point in $\mathrm{Tot} X^\bullet$ which is non-empty. If $X^\bullet$ is not pointed $\mathrm{Tot} X^\bullet$ may be empty. However, let b be in $\pi^0 \pi_0 X^\bullet$. The set $\pi^0 \pi_0 X^\bullet$ identifies with the image of $\pi_0 \mathrm{Tot}_1 X^\bullet$ in $\pi_0 \mathrm{Tot}_0 X^\bullet$. Therefore b is the class of a 0-simplex of $\mathrm{Tot}_0 X^\bullet$ which lifts to $\mathrm{Tot}_1 X^\bullet$. More generally one can ask for conditions ensuring that this 0-simplex lifts to $\mathrm{Tot} X^\bullet$.

Theorem 8.2.2 [Bo4]. *Let $X^\bullet$ be an $\mathbb{F}_p$-affine-like cosimplicial space. Let b be in $\pi^0 \pi_0 X^\bullet$. If $\pi^s \pi_{s-1} X^\bullet$ is trivial for all $s \geq 2$, then b lifts to $\mathrm{Tot} X^\bullet$.*

We are using the last criterion in ([**Bo4**] 6.1), noting that there are no Whitehead products in our affine-like setting. This allows to choose a base point in $X^\bullet$. This result was first stated in a letter of Bousfield to Neisendorfer. For a proof of Theorem 8.2.2 see also the appendix B of [**L4**].

We shall need a second theorem telling us when $\mathrm{Tot} X^\bullet$ is connected. This result is more classical.

Theorem 8.2.3 [BK2]. *Let $X^\bullet$ be a fibrant pointed cosimplicial space. Then $\mathrm{Tot}\, X^\bullet$ is connected as soon as $\pi^s \pi_s X^\bullet = 0$ for all $s \geq 0$.*

We note that this theorem makes sense even if $X^\bullet$ is not group-like. In this case we refer to **[BK2]** for the definition of $\pi^0 \pi_0 X^\bullet$ and $\pi^1 \pi_1 X^\bullet$.

8.3. Resolutions of spaces and mapping spaces

Let us come back to our case. Let R be a commutative, unital ring. On the category of simplicial sets $s\mathcal{S}$, we consider the 'monad' (R, Φ, Ψ) defined as follows.

The forgetful functor $\mathcal{O}$ from $s\mathcal{M}od_R$ (simplicial R-modules) to $s\mathcal{S}$ has a left adjoint $\mathcal{F}$. The composite functor $R = \mathcal{O} \circ \mathcal{F} : s\mathcal{S} \to s\mathcal{S}$ comes with two natural transformations

the unit of the adjunction $\Phi : \mathrm{Id} \to R$ and

$\Psi : R^2 \to R$ which is equal to $\mathcal{F}$ applied to the adjoint of Id_R. The space RX is the free simplicial R-module generated by the space X; in simplicial degree n it has the set X_n as basis. The following diagrams commute for any space X

$$
\begin{array}{ccc}
R^3 X & \xrightarrow{\;R\Psi_X\;} & R^2 X \\
{\scriptstyle \Psi_{RX}}\downarrow & & \downarrow{\scriptstyle \Psi_X} \\
R^2 X & \xrightarrow{\;\Psi_X\;} & RX
\end{array}
$$

and

$$
\begin{array}{ccc}
RX & \xrightarrow{\;\Phi_{RX}\;} & R^2 X \\
{\scriptstyle R\Phi_X}\downarrow & \searrow{\scriptstyle \mathrm{Id}} & \downarrow{\scriptstyle \Psi_X} \\
R^2 X & \xrightarrow{\;\Psi\;} & RX
\end{array}
$$

The set (R, Φ, Ψ) is called a monad. It is standard that the following is a well defined cosimplicial space:

$$
\begin{cases}
(R^\bullet X)^n = R^{n+1} X \,; \\
d^i : (R^\bullet X)^{n-1} \to (R^\bullet X)^n & \text{is equal to} \quad R^i \Phi_{R^{n-i} X} \,; \\
s^i : (R^\bullet X)^{n+1} \to (R^\bullet X)^n & \text{is equal to} \quad R^i \Psi_{R^{n-i} X} \,.
\end{cases}
$$

The cosimplicial space $R^\bullet X$ is coaugmented over X: the map $\Phi :$ $X \to RX = (R^\bullet X)^0$ is such that $d^0 \circ \Phi = d^1 \circ \Phi$. More generally $d_{i_n} \circ \ldots \circ d_{i_1} \circ \Phi$ does not depend on $i_1, \ldots, i_n$. This defines a map from the constant cosimplicial space X to $R^\bullet X$. Therefore this defines a map $d : \mathrm{Tot}\, X = X \to \mathrm{Tot}\, R^\bullet X$.

The functor

$$X \mapsto \mathrm{Tot}\, R^\bullet X$$

is known as the Bousfield-Kan R-*completion* of X. The space $\mathrm{Tot}\, R^\bullet X$ will be denoted by $R_\infty X$.

The cosimplicial space $R^\bullet X$ is group-like, hence fibrant. Let Y be a connected pointed space. Then

Proposition 8.3.2 [BK2]. *The canonical map* $\mathrm{Tot}_s R^\bullet Y \to \mathrm{Tot}_{s-1} R^\bullet Y$ *is a principal fibration.*

Given a pointed tower of fibrations $\{X_s; f_s\}$, $s \in \mathbb{N}$, there is a homotopy spectral sequence **[BK2]** converging to the homotopy of $\lim_s X_s$ under nice hypotheses. We are interested in the following two cases:

(i) $X_s = \mathrm{Tot}_s R^\bullet Y$ for some space Y;

(ii) $X_s = \mathrm{map}(X, \mathrm{Tot}_s R^\bullet Y) = \mathrm{Tot}_s \mathrm{map}(X, R^\bullet Y)$. In the first case, we choose any base point in Y, this gives a base point in the tower of fibrations. In the second case, any map $f : X \to R_\infty Y$ induces a base point in the tower of fibrations. And in order to apply Proposition 8.3.2 one replaces $\mathrm{map}(X, \mathrm{Tot}_s \mathbb{F}_p^\bullet Y)$ by the connected component of this base point (see 8.4 for the definition).

Let $X^\bullet$ be a fibrant and pointed cosimplicial space. Bousfield and Kan have shown that the $E_2^{s,t}$-term, defined for $t \geq s \geq 0$, of the homotopy spectral sequence of the tower of fibrations

$$\to \mathrm{Tot}_s X^\bullet \to \mathrm{Tot}_{s-1} X^\bullet \to \ldots ,$$

is isomorphic to $\pi^s \pi_t X^\bullet$.

At this point, we warn the reader that the choice of the monad on the category $s\mathcal{S}$ is not quite the one of **[BK2]**. However, it turns out that the two monads give the same total space (see also **[L4]**).

8.4. Proof of Theorem 8.1.1

We now specialize to the case $R = \mathbb{F}_p$ in the preceding constructions.

Consider first the case where we choose the trivial map as base point in the cosimplicial space $\mathrm{map}(X, \mathbb{F}_p^\bullet Y)$ and, therefore, in the tower $\{\mathrm{Tot}_s\,\mathrm{map}(X, \mathbb{F}_p^\bullet Y)\}$, $s \in \mathbb{N}$. Assume also that H^*X and H^*Y are of finite type and that X and Y are connected. Then following [BK1]:

Proposition 8.4.1. *With the preceding choice of base points*

$$\pi^s \pi_t\,\mathrm{map}(X, \mathbb{F}_p^\bullet Y) \cong \mathrm{Ext}_{\mathcal{K}}^s(H^*Y, \Sigma^t(H^*X)^+)$$

if $t \geq 1$, $s \geq 0$ *and*

$$\pi^0 \pi_0\,\mathrm{map}(X, \mathbb{F}_p^\bullet Y) \cong \mathrm{Hom}_{\mathcal{K}}(H^*Y, H^*X).$$

Proof. Recall the following theorem. Let E be a graded $\mathbb{F}_p$-vector space of finite type ($\dim_{\mathbb{F}_p} E_n < +\infty$ for each n). Denote by $K(E)$ the product $\prod_n K(E_n, n)$.

Recall that the functor $\mathcal{G}$ is left adjoint to the forgetful functor $\mathcal{O}$ from $\mathcal{K}$ to $\mathcal{E}_{gr}$ and that the functor G is the composite $\mathcal{G} \circ \mathcal{O}$. Then

Theorem (Cartan, Serre) [C][S1]. *The mod p cohomology $H^*K(E)$ is isomorphic to* $\mathcal{G}(E^*) \cong \otimes_n U(E_n^* \otimes F(n))$.

See Section 3.8 for the definition of the functor U.

It follows easily from this theorem that for any connected space X such that H^*X is of finite type one has a bijection

$$[X, K(E)] \cong \mathrm{Hom}_{\mathcal{K}}(H^*K(E), H^*X) \cong \mathrm{Hom}_{\mathcal{K}}(H^*\mathcal{G}(E^*), H^*X).$$

The space $\mathbb{F}_p Y$ is homotopically equivalent to

$$K(\mathcal{O}(H^*Y)) = \prod_{n \geq 0} K(H_n Y, n).$$

Note that as Y is not pointed, we do not use reduced cohomology. Observe also that if H^*Y is of finite type so is $H^*\mathbb{F}_p Y$.

Now let us compute $\pi_t\,\mathrm{map}(X, (\mathbb{F}_p^\bullet Y)^s)$, $t > 0$. It is isomorphic to the subset of elements of $[S^t \times X, (\mathbb{F}_p^\bullet Y)^s]$ which are mapped onto the homotopy class of the trivial map in $[X, (\mathbb{F}_p^\bullet Y)^s]$ by composition with the inclusion $* \times X \hookrightarrow S^t \times X$. Therefore, this is the subset of

$$\text{Hom}_{\mathcal{K}}((G_\bullet(H^*Y))_s, H^*S^t \otimes H^*X)$$

which, by composing with the projection $H^*S^t \otimes H^*X \to H^*X$, is mapped onto the trivial map in

$$\text{Hom}_{\mathcal{K}}((G_\bullet(H^*Y))_s, H^*X).$$

This is nothing other than $\text{Hom}_{\mathcal{K}}((G_\bullet(H^*Y))_s, \Sigma^t H^*X^+)$, where $\Sigma^t H^*X^+$ has the same meaning here as in (7.1.7).

Obviously, all this could be expressed in purely homotopic terms. The case $t = 0$ is easily done and one finds

$$\pi_0 \text{map}(X, \mathbb{F}_p^\bullet Y) \cong \text{Hom}_{\mathcal{K}}((G_\bullet(H^*Y))_s, H^*X).$$

Now if $s \geq 1$, by the very definition

$$\pi^s \pi_t \text{map}(X, \mathbb{F}_p^\bullet Y) \cong \text{Ext}_{\mathcal{K}}^s(H^*Y, \Sigma^t H^*X^+).$$

If $s = 0$, we need a special argument. The set $\pi^0 \pi_0 \text{map}(X, \mathbb{F}_p^\bullet Y)$ identifies with the equalizer of the two maps

$$d^0, d^1 : \text{Hom}_{\mathcal{K}}((G_\bullet(H^*Y))_0, H^*X) \to \text{Hom}_{\mathcal{K}}((G_\bullet(H^*Y))_1, H^*X).$$

And it is clearly enough to check that, for any unstable $\mathcal{A}$-algebra L, the coequalizer of the two maps d_0 and d_1 from $(G_\bullet(L))_1$ to $(G_\bullet(L))_0$ is L itself. This follows from 7.1.1 (i) and (iii). The result can be shown by direct computation using the following formulas describing d_0 and d_1. Denote by y_i generic elements of H^*Y. Elements of $(G_\bullet(H^*Y))_0 = G(H^*Y)$ are sums of monomials $\prod_i \theta_i[y_i], \theta_i \in \mathcal{A}$; elements of $(G_\bullet(H^*Y))_1 = G^2(H^*Y)$ are sums of monomials $\prod_j \nu_j[\prod_i \theta_{i,j}[y_{i,j}]], \nu_j, \theta_{i,j} \in \mathcal{A}$. The formulas describing d_0 and d_1 are:

$$d_0\Big(\prod_j \nu_j\big[\prod_i \theta_{i,j}[y_{i,j}]\big]\Big) = \prod_{i,j} \nu_j \theta_{i,j}[y_{i,j}],$$

$$d_1\Big(\prod_j \nu_j\big[\prod_i \theta_{i,j}[y_{i,j}]\big]\Big) = \prod_{i,j} \nu_j\big[\prod_i \theta_{i,j} y_{i,j}\big].$$

Now let φ be an element of $\text{Hom}_{\mathcal{K}}(H^*Y, H^*X)$. We define a cosimplicial subspace, $\text{map}(X, \mathbb{F}_p^\bullet Y)_\varphi \subset \text{map}(X, \mathbb{F}_p^\bullet Y)$ by the following requirement: $X \times \Delta[n] \to (\mathbb{F}_p^{\bullet} Y)^m$ is an n simplex of $\mathrm{map}(X, (\mathbb{F}_p^{\bullet} Y)_{\varphi}^m)$ if and only if the induced map in cohomology

$$(G_{\bullet}(H^*Y))_m = G^{m+1}(H^*Y) \to H^*X$$

factors as $\varphi \circ d^*$ where $d : Y \to \mathbb{F}_p^{\,s+1} Y$ is induced by the coaugmentation.

Let ε be the trivial map in $\mathrm{Hom}_{\mathcal{K}}(H^*Y, H^*X)$ The cosimplicial space $\mathrm{map}(X, \mathbb{F}_p^{\bullet} Y)_{\varepsilon}$ is pointed and $\pi^s \pi_t \mathrm{map}(X, \mathbb{F}_p^{\bullet} Y)_{\varepsilon}$ agrees with the formulas (8.4.1) except when $t = 0$, in which case:

$$\pi^0 \pi_0 \mathrm{map}(X, \mathbb{F}_p^{\bullet} Y)_{\varepsilon} = \{\varepsilon\} \,.$$

Theorem 8.4.2. *The subspace of* $\mathrm{map}(BV, \mathbb{F}_{p\infty} Y)$ *of maps inducing the trivial map in cohomology is connected.*

Proof. We are looking at $\mathrm{Tot}\,\mathrm{map}(BV, \mathbb{F}_p^{\bullet} Y)_{\varepsilon}$. By Theorem 8.2.3, it is enough to show that $\pi^s \pi_s \mathrm{map}(BV, \mathbb{F}_p^{\bullet} Y)_{\varepsilon}$ is trivial if $s \geq 1$, thus, that $\mathrm{Ext}_{\mathcal{K}}^s(H^*Y, \Sigma^s H^*V^+)$ (see 7.1.7) is trivial for any $s \geq 1$. This follows from Theorem 7.8.3.

Next let φ in $\mathrm{Hom}_{\mathcal{K}}(H^*Y, H^*V)$ be any map. The cosimplicial space $\mathrm{map}(BV, \mathbb{F}_p^{\bullet} Y)_{\varphi}$ is affine-like under the $\mathbb{F}_p$-vector space-like cosimplicial space $\mathrm{map}(BV, \mathbb{F}_p^{\bullet} Y)_{\varepsilon}$. But, *a priori*, it is not pointed. However, one can apply Theorem 8.2.2 in order to show that $\mathrm{Tot}\,\mathrm{map}(BV, \mathbb{F}_p^{\bullet} Y)_{\varphi}$ is non empty. We need the following

Proposition 8.4.3 [Bo4]. *For* $t \geq 1$ *and* $s \geq 0$

$$\pi^s \pi_t \mathrm{map}(X, \mathbb{F}_p^{\bullet} Y)_{\varphi} \cong \mathrm{Der}_{\mathcal{K}/H^*X}^s(H^*Y, \Sigma^t H^*X) \,.$$

Note that we do not need a base point to define $\pi^s \pi_t \mathrm{map}(X, \mathbb{F}_p^{\bullet} Y)_{\varphi}$. This proposition was first stated in the letter of Bousfield mentioned above.

Obviously $\pi^0 \pi_0 \mathrm{map}(X, \mathbb{F}_p^{\bullet} Y)_{\varphi} = \{\varphi\}$. In our case

$$\pi^s \pi_t \mathrm{map}(BV, \mathbb{F}_p^{\bullet} Y)_{\varphi} = \mathrm{Der}_{\mathcal{K}/H^*V}^s(H^*Y, \Sigma^t H^*V)$$

is trivial as soon as $t - s \leq 0$. Therefore Theorem 8.2.2 implies that $\mathrm{Tot}\,\mathrm{map}(BV, \mathbb{F}_p^{\bullet} Y)_{\varphi}$ is non empty. This is the space of maps $f : BV \to \mathbb{F}_{p\infty} Y$ factorizing through φ in cohomology:

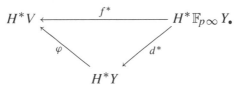

where $d : Y \rightarrow \mathbb{F}_{p\infty} Y$ is the map induced by the coaugmentation. An element in $\text{Tot map}(BV, \mathbb{F}_p^{\bullet} Y)_\varphi \cong \text{map}(BV, \mathbb{F}_{p\infty} Y)_\varphi$ defines base points in the tower of fibrations

$$\rightarrow \text{Tot}_s \text{map}(BV, \mathbb{F}_p^{\bullet} Y)_\varphi \rightarrow \text{Tot}_{s-1} \text{map}(BV, \mathbb{F}_p^{\bullet} Y)_\varphi \rightarrow .$$

The connectivity lemma 8.2.3 shows that $\text{Tot map}(BV, \mathbb{F}_p^{\bullet} Y)_\varphi$ is connected as soon as the $E_2^{s,s}$ term vanishes for $s \geq 0$. But $E_2^{0,0} \cong \{\varphi\}$ and $E_2^{s,s} \cong \text{Der}^s_{\mathcal{K}/H^*V}(H^*Y, \Sigma^s H^*V)$ for $s \geq 1$ which is trivial by Theorem 7.8.3. Therefore

Theorem 8.4.4. *Let φ be in $\text{Hom}_{\mathcal{K}}(H^*Y, H^*V)$. The subspace of maps of $\text{map}(BV, \mathbb{F}_{p\infty} Y)$ inducing $\varphi \circ d^*$ is non-empty and connected.*

It follows from the comments following (8.2.1*) and (8.4.1) that $[BV, \mathbb{F}_{p\infty} Y]$ is in bijection with $\text{Hom}_{\mathcal{K}}(H^*Y, H^*V)$. It remains to compare $[BV, Y]$ to $[BV, \mathbb{F}_{p\infty} Y]$.

Recall that a space Y is said to be virtually nilpotent if the fundamental group contains a subgroup of finite index that acts nilpotently on higher homotopy groups. In particular a space with finite fundamental group is virtually nilpotent. For a virtually nilpotent space Y the map $d : Y \rightarrow \mathbb{F}_{p\infty} Y$ is an $H_*(-, \mathbb{F}_p)$-Bousfield localization and induces an isomorphism in mod p cohomology **[BK2][Bo2]**. The arithmetic square of Dror, Dwyer and Kan **[DDK]** allows us to conclude that for Y virtually nilpotent there is a fibre square up to homotopy:

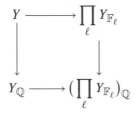

where Y_R denotes the $H_*(-, R)$-Bousfield localization of Y. Suppose that the spaces are pointed and apply the functor $\mathrm{map}_*(BV, -)$. We get a fibre square up to homotopy of mapping spaces. As the spaces $\mathrm{map}_*(BV, Y_{\mathbb{F}_\ell})$ ($\ell \neq p$), $\mathrm{map}(BV, Y_{\mathbb{Q}})$ and $\mathrm{map}_*(BV, (\prod_\ell Y_{\mathbb{F}_\ell})_{\mathbb{Q}})$ are contractible we have

$$\mathrm{map}_*(BV, Y) \cong \mathrm{map}_*(BV, \mathbb{F}_{p\infty} Y).$$

This argument is given by Miller in [Ml2]. In particular, we have

$$[BV, Y]_* \cong [BV, \mathbb{F}_{p\infty} Y]_*.$$

If $\pi_1 Y$ is trivial we are done, otherwise we need to analyze the diagram:

$$
\begin{array}{ccc}
[BV, Y]_* & \xrightarrow{\ d\ } & [BV, \mathbb{F}_{p\infty} Y]_* \\
\downarrow & & \downarrow \\
[BV, Y] & \longrightarrow & [BV, \mathbb{F}_{p\infty} Y]
\end{array}
$$

This is done in [L1] by Lannes. To prove that the quotient map is a bijection it is enough to know that $\pi_1 Y \to \pi_1 \mathbb{F}_{p\infty} Y$ is surjective. This is true if $\pi_1 Y$ is finite [BK2][Bo2].

To conclude we note that Lannes' theorem is stronger than (8.1.1). He proves that $[BV, Y] \to \mathrm{Hom}_{\mathcal{K}}(H^*Y, H^*V)$ is a homeomorphism for the natural topology on each side [L4].

8.5. Comments about the connected components of the total space

In this section we give some information about $\pi_0 \mathbb{F}_{p\infty} X$ which is going to be useful later (see also [L4]). The coaugmentation Φ induces a map $\pi_0 X \to \pi_0 \mathbb{F}_{p\infty} X$. On the other hand there is a map $\pi_0 \mathbb{F}_{p\infty} X \to \pi_0 X$ which is induced by the canonical map $X \to \pi_0 X$. The composite is the identity. Therefore, the map $\pi_0 X \to \pi_0 \mathbb{F}_{p\infty} X$ is injective. In fact:

Proposition 8.5. *The canonical map $\pi_0 X \to \pi_0 \mathbb{F}_{p\infty} X$ is bijective.*

This a consequence of (8.2.3) and is shown as follows.

Let x be a 0-simplex in X representing an element ω of $\pi_0 X$. Consider the cosimplicial subspace $(\mathbb{F}_p^{\bullet} X)_x$. It is the subcosimplicial subspace of $\mathbb{F}_p^{\bullet} X$ consisting of elements of $\mathbb{F}_p^{\bullet} X$ which are in the connected component of some coface of x. The space $(\mathbb{F}_{p\infty} X)_x$ identifies with the subspace of $\mathbb{F}_{p\infty}$ mapped to ω by the composite map $\mathbb{F}_{p\infty} X \to \pi_0 \mathbb{F}_{p\infty} X \to \pi_0 X$. Thus it is enough to check that $(\mathbb{F}_{p\infty} X)_x$ is connected. Theorem 8.2.3 and an easy extension of Proposition 8.4.1 show that it is enough to prove the triviality of $\text{Ext}^s_{\mathcal{K}_a}(H^* X, \Sigma^s \mathbb{F}_p^{+})$ for $s \geq 1$ (the augmentation of $H^* X$ being determined by ω). This follows from 7.4.1.

8.6. H. Miller's theorem and a converse

In this section we prove

Theorem 8.6.1 [Ml2][LS2]. *Let X be a connected nilpotent space such that $H^* X$ is of finite type and $\pi_1 X$ is finite. The following two conditions are equivalent:*

(i) *the canonical map $H^* X \to T H^* X$ is an isomorphism in degrees less or equal to k,*

(ii) *the space of pointed maps $\text{map}_*(B\mathbb{Z}/p, X)$ is k-connected.*

Here is an immediate consequence (using Theorem 7.2.1).

Corollary 8.6.2. *Let X be a connected nilpotent space such that $H^* X$ is of finite type and $\pi_1 X$ is finite. Then the following two conditions are equivalent:*

(i) *$H^* X$ is locally finite;*

(ii) *the space of pointed maps $\text{map}_*(B\mathbb{Z}/p, X)$ is contractible.*

The implication in the corollary (i) $\Rightarrow$ (ii) is the extension of Miller's theorem given in **[LS1]** except that we assume here that $H^* X$ is of finite type. We are going to prove Theorem 8.6.1 using the techniques of the preceding chapters. Then we shall relate it to the material Chapter 6 using the Eilenberg-Moore spectral sequence. This will give us opportunities to give applications of the preceding algebraic machinery to the Eilenberg-Moore spectral sequence.

Proof of Theorem 8.6.1.

The implication (i) $\Rightarrow$ (ii) has essentially been done in the preceding chapters. This is proved as follows.

Consider the evaluation map at the base point of $B\mathbb{Z}/p$

$$\omega : \text{map}(B\mathbb{Z}/p, X) \to X.$$

It is enough to show that this map is k-connected (bijective on the π_i's for $i < k$ and surjective on π_k). We use the following version of the mapping lemma [**BK2**].

Lemma 8.6.3. *Let* $\{X_s\}$, $s \in \mathbb{N}$, *be a tower of pointed fibrations such that in the homotopy spectral sequence* $E_2^{s,s} = *$ *for* $s \geq 0$ *and let* $f_s : \{X_s\} \to \{Y_s\}$ *be a map of tower of fibrations inducing an isomorphism for* $k \geq t - s \geq 0$ *and an epimorphism for* $t - s = k$ *on the* $E_2^{s,t}$ *-terms of the homotopy spectral sequences. Then the induced map* $f : \lim_s X_s \to \lim_s Y_s$ *is* k-*connected.*

The case we are interested in is where

$$
\begin{aligned}
\{X_s\}_{s \in \mathbb{N}} &= \{\text{Tot}_s(\text{map}(B\mathbb{Z}/p, \mathbb{F}_p^\bullet X))\}_{s \in \mathbb{N}} \\
&= \{\text{map}(B\mathbb{Z}/p, \text{Tot}_s \mathbb{F}_p^\bullet X)\}_{s \in \mathbb{N}}
\end{aligned}
$$

and $\{Y_s\} = \{\text{Tot}_s \mathbb{F}_p^\bullet X\}$, $s \in \mathbb{N}$. The map f is the evaluation at the base point of $B\mathbb{Z}/p$. The fact that the $E_2^{s,s}$ term of the homotopy spectral sequence of the tower $\{X_s\}$ (as well as for the tower $\{Y_s\}$) is trivial for $s \geq 1$ was proved in Section 8.4. The fact that $E_2^{0,0}$ is trivial follows directly from the hypothesis that $H^*X \to TH^*X$ is an isomorphism in degree zero.

Consider the maps induced on the $E_2^{s,t}$-terms. These are:

$$\text{Ext}_{\mathcal{K}}^s(H^*X, H^*S^t \otimes H^*\mathbb{Z}/p) \to \text{Ext}_{\mathcal{K}}^s(H^*X, H^*S^t).$$

We have to show these are isomorphisms for $0 \leq t - s \leq k - 1$ and epimorphisms for $t - s = k$. Note that the left term is isomorphic to $\text{Ext}_{\mathcal{K}}^s(TH^*X, H^*S^t)$, the map is induced by the inclusion $H^*X \to TH^*X$.

Proposition 8.6.4. *Let* $K \to L$ *be a map of connected unstable* $\mathcal{A}$-*algebras which is an isomorphism in degrees less than or equal to* $k - 1$, *a monomorphism in degree* k. *Then, there exists a free* *resolution $K_\bullet$ (resp. $L_\bullet$) of K (resp. of L) and a map $f_\bullet : K_\bullet \to L_\bullet$ such that:*

(i) *$f_n : K_n \to L_n$ is an isomorphism in degrees less than or equal to $k - 1 + n$,*

(ii) *$f_n : K_n \to L_n$ is an monomorphism in degree $k + n$.*

This is proved using a step by step construction. As an immediate consequence we get taking $K = H^*X$ and $L = TH^*X$ that

$$\mathrm{Ext}^s_{\mathcal{K}}(L, \Sigma^t \mathbb{F}_p{}^+) \to \mathrm{Ext}^s_{\mathcal{K}}(K, \Sigma^t \mathbb{F}_p{}^+)$$

is an isomorphism if $0 \le t - s \le k - 1$, an epimorphism if $t - s = k$. This completes the proof of (i) $\Rightarrow$ (ii).

The reverse implication of Theorem 8.6.1 will be proved by using the following

Lemma 8.6.5. *Let K be a connected unstable algebra and k be an integer. The following conditions are equivalent:*

(i) $\mathrm{Ext}^s_{\mathcal{K}}(K, H^*S^t \otimes H^*\mathbb{Z}/p) \cong \mathrm{Ext}^s_{\mathcal{K}}(K, H^*S^t)$ *for $0 \le t - s \le k$,*

(ii) $\mathrm{Hom}_{\mathcal{K}}(K, H^*S^t \otimes H^*\mathbb{Z}/p) \cong \mathrm{Hom}_{\mathcal{K}}(K, H^*S^t)$ *for $0 \le t \le k$,*

(iii) $K \to TK$ *is an isomorphism in degrees less or equal to k.*

Of course, (i) $\Rightarrow$ (ii) is also trivial, and (iii) $\Rightarrow$ (i) follows from Proposition 8.6.4.

Let us show that (ii) $\Rightarrow$ (iii). By condition (ii) the map $K \to TK$ induces a bijection from $\mathrm{Hom}_{\mathcal{K}}(TK, H^*S^t)$ to $\mathrm{Hom}_{\mathcal{K}}(K, H^*S^t)$, for $0 \le t \le k$. By Theorem 3.9.6, $\mathrm{Hom}_{\mathcal{U}}(TK, H^*S^t)$ is isomorphic to $\mathrm{Hom}_{\mathcal{U}}(K, H^*S^t)$ for $0 \le t \le k$. Therefore, $\mathrm{Hom}_{\mathcal{U}}(TK, J(t))$ is isomorphic to $\mathrm{Hom}_{\mathcal{U}}(K, J(t))$ for $0 \le t \le k$. Consequently, $(TK)^t$ is isomorphic to K^t for $0 \le t \le k$. The result follows.

Proof of (ii) $\Rightarrow$ (i) in Theorem 8.6.1.

Suppose that $\pi_0 \mathrm{map}_*(B\mathbb{Z}/p, \mathbb{F}_{p\infty} X) \cong \pi_0 \mathrm{map}_*(B\mathbb{Z}/p, X)$ is trivial. By Theorem 8.1.1 we know that $\mathrm{Hom}_{\mathcal{K}}(H^*X, H^*\mathbb{Z}/p)$ is trivial. By Lemma 8.6.5 we are done.

Suppose now that $\pi_1 \operatorname{map}_*(B\mathbb{Z}/p, X)$ is trivial. We claim that the map

$$\operatorname{Hom}_{\mathcal{K}}(H^*X, H^*S^1 \otimes H^*\mathbb{Z}/p) \cong \operatorname{Hom}_{\mathcal{K}}(TH^*X, H^*S^1)$$
$$\to \operatorname{Hom}_{\mathcal{K}}(H^*X, H^*S^1)$$

is a bijection. Clearly this is a surjection. It is an injection, for an element z in the kernel is an infinite cycle in the homotopy spectral sequence for $\operatorname{map}(B\mathbb{Z}/p, X)$. Hence it yields an element in $\pi_1 \operatorname{map}(B\mathbb{Z}/p, \mathbb{F}_{p\infty} X)$ which is killed by the evaluation map. Therefore, this element comes from $\pi_1 \operatorname{map}_*(B\mathbb{Z}/p, \mathbb{F}_{p\infty} X)$ and must be zero. Applying Lemma 8.6.5 we conclude that the map $H^*X \to TH^*X$ is an isomorphism in degree 1. Observe that we have used the bijection

$$\operatorname{Hom}_{\mathcal{K}}(H^*X, \Sigma H^*\mathbb{Z}/p^+) \cong \operatorname{Hom}_{\mathcal{K}}(H^*X, H^*S^1 \otimes H^*\mathbb{Z}/p)$$

which is forced by the hypothesis

$$\operatorname{Hom}_{\mathcal{K}}(H^*X, H^*\mathbb{Z}/p) = \{*\}.$$

For $k > 1$ one proceeds by induction.

We finish this section by recalling Miller's original strategy to prove that $\operatorname{Ext}^s_{\mathcal{K}}(H^*X, \Sigma^t \widetilde{H}^*\mathbb{Z}/p^+)$ is trivial as soon as H^*X is finite. Using the spectral sequence of (7.5) one is reduced to computing $\operatorname{Ext}^u_{\mathcal{U}}(L_v^S \Omega Q(H^*X), \Sigma^{t-1} \widetilde{H}^*\mathbb{Z}/p)$. A careful analysis of André-Quillen cohomology show that the left terms are finite. This is proved by reducing to the case $t = 0$ by using another spectral sequence. One has to check that the derived functors of Ω^t preserve finiteness and one can conclude.

8.7. Applications of the Eilenberg-Moore spectral sequence

We first recall from [R] and [Sm] some facts about the Eilenberg-Moore spectral sequence. Consider the case of the fibre square

The Eilenberg-Moore spectral sequence is a second quadrant cohomological spectral sequence which converges strongly towards $H^*\Omega X$ as soon as $\pi_1 X = 0$ (for a more general result see [**D**]).

Its $E_2^{s,*}$-term identifies, as an $\mathcal{A}$-module, with $\mathrm{Tor}^s_{H^*X}(\mathbb{F}_p, \mathbb{F}_p)$ provided with the $\mathcal{A}$-module structure defined by the Cartan formula. The columns $E_r^{s,*}$, $s \leq 0$, $2 \leq r \leq \infty$ are unstable $\mathcal{A}$-modules and it is convenient to interpret the differentials as $\mathcal{A}$-linear *of degree zero* from $E_r^{s,*}$ to $\Sigma^{r-1} E_r^{s,*}$:

$$d_r : E_r^{s,*} \to \Sigma^{r-1} E_r^{s+r,*} .$$

There is a convergent filtration by $\mathcal{A}$-modules on $H^*\Omega X$:

$$H^*\Omega X \supset ...F_s \supset F_{s+1} \supset ... \supset F_{-1} \supset F_0 = \mathbb{F}_p ,$$

such that $\sum^{-s} F_s/F_{s+1} \cong E_\infty^{s,*}$.

The kernel K of the edge homomorphism $E_2^{-1,*} \to \Sigma H^*\Omega X$ has a convergent filtration: $K \supset ...K_s \supset K_{s+1} \supset ... \supset K_{-1} = 0$ such that $\sum^{-s-1} K_s/K_{s+1} \cong d_{-s} E_{-s}^{s+1,*}$.

Let X be a 1-connected space such that H^*X is of finite type and let $\sigma : \Sigma\Omega X \to X$ be the evaluation map.

Theorem 8.7.1. *Assume that the mod p cohomology of X is in $\overline{\mathcal{N}il}_\ell$, for some $\ell \geq 1$. Then $H^*\Omega X$ is in $\overline{\mathcal{N}il}_{\ell-1}$. Moreover*

(i) *if $\ell \geq 2$, σ^* has kernel and cokernel in $\overline{\mathcal{N}il}_{\ell+1}$;*

(ii) *if $\ell = 1$, σ^* has kernel in $\overline{\mathcal{N}il}_2$.*

Finally if the mod p cohomology of X is in $\overline{\mathcal{N}il}_0 \cong \mathcal{U}$ and if QH^*X is in $\overline{\mathcal{N}il}_k$, $k \geq 1$, then the map $QH^*X \to \Sigma H^*\Omega X$ has kernel in $\overline{\mathcal{N}il}_{k+1}$.

Corollary 8.7.2. *Keep the same hypothesis on X, and assume moreover that H^*X is locally finite (as an $\mathcal{A}$-module). Then so is $H^*\Omega X$.*

Proof. This is because $\cap_\ell \overline{\mathcal{N}il}_\ell = \mathcal{B}$.

Before proving Theorem 8.7.1 we state.

Proposition 8.7.3. *Let X be a homotopy associative H-space such that H^*X is of finite type and let $\widetilde{X} \xrightarrow{\pi} X$ be a covering. If H^*X belongs to $\overline{\mathcal{N}il}_k$, for some $k \geq 1$, then the kernel and cokernel of π^* are in $\overline{\mathcal{N}il}_{k+1}$.*

Using Corollary 8.7.2 and Proposition 8.7.3 one gets

Corollary 8.7.4. *Let X be a 1-connected space such that H^*X is of finite type. Then, if H^*X is locally finite as an A-module, so is $H^*\Omega^n X$ for any n.*

Proof. One applies 8.7.3 to ΩX and the universal cover $\widetilde{\Omega}X$. Then one applies 8.7.2 to $\widetilde{\Omega}X$. We refer to [**LS1**] for analogous (in fact more general) homological results.

Proof of Theorem 8.7.1.

Consider statement (i) of 8.7.1.

Lemma 8.7.5. *Let M be an unstable module. If M is in $\overline{Nil}_\ell$, $\ell \geq 1$, and j-connected with $j \leq \ell - 1$. Then $M^{\otimes n}$ is in $\overline{Nil}_{\ell+(j+1)(n-1)}$.*

The proof of this lemma is left to the reader.

We have $H^*X \in \overline{Nil}_\ell$, $\ell \geq 2$, and $\widetilde{H}^*X$, 1-connected. Thus, $E_2^{s,*}$, which is a quotient of $\widetilde{H}^*X^{\otimes -s}$, is in $\overline{Nil}_{\ell-2(s+1)}$. Consequently, $E_\infty^{s,*}$ is also in $\overline{Nil}_{\ell-2(s+1)}$. Hence $\Sigma^{-s}F_s/F_{s+1}$ is also in $\overline{Nil}_{\ell-2(s+1)}$ and (using Theorem 6.2.2) F_s/F_{s+1} is in $\overline{Nil}_{\ell-s-2}$. It follows that $H^*\Omega X$ is in $\overline{Nil}_{\ell-1}$ and that F_{-2} is in $\overline{Nil}_\ell$. Thus the inclusion $F_{-1} \hookrightarrow H^*\Omega X$ has its cokernel in $\overline{Nil}_\ell$.

Next look at the kernel K of $E_2^{-1,*} \to E_\infty^{s-1,*} = F_{-1}$. As above, we know that $E_{-s}^{s-1,*}$ is in $\overline{Nil}_{\ell-2s}$, therefore so is $\Sigma^{-s+1}K_s/K_{s+1}$. Thus (using Theorem 6.2.2) K_s/K_{s+1} is in $\overline{Nil}_{\ell-s-1}$ and K is in $\overline{Nil}_{\ell+1}$. Now, the map $\sigma^* : H^*X \to \Sigma H^*\Omega X$ factors as

$$
\begin{array}{ccc}
H^*X & \longrightarrow & \Sigma H^*\Omega X \\
\downarrow & & \uparrow \\
QH^*X = E_2^{-1,*} & \longrightarrow & E_\infty^{-1,*}
\end{array}
$$

The result follows by observing that the kernel of $H^*X \to QH^*X$ is in $\overline{Nil}_{\ell+2}$, as it is a quotient of $(\widetilde{H}^*X)^{\otimes 2}$.

Consider now statement (ii) of Theorem 8.7.1. The same argument as above applies, except that we have no control on the cokernel of $F_{-1} \hookrightarrow H^*\Omega X$ which, in general, does not belong to $\overline{Nil}_1$.

Statement (iii) is the hardest one. This is because we cannot use an elementary lemma of the type of 8.7.5. However, we know from Theorem 6.4.1 that for any unstable $\mathcal{A}$-algebra K $\mathrm{Tor}_K^s(\mathbb{F}_p, \mathbb{F}_p)$ is in $\overline{Nil}_{-s}$. More generally.

Lemma 8.7.6. *Let K be an unstable $\mathcal{A}$-algebra such that $Q(K)$ is in $\overline{Nil}_\ell$, $\ell \geq 0$. Then $\mathrm{Tor}_K^s(\mathbb{F}_p, \mathbb{F}_p)$ is in $\overline{Nil}_{\ell-s-1}$.*

See also [**Sc1**].

Proof. It is enough to check that, for any V,

$$\mathrm{Tor}_K^s(\mathbb{F}_p, \mathbb{F}_p) \to T_V(\mathrm{Tor}_K^s(\mathbb{F}_p, \mathbb{F}_p)) = \mathrm{Tor}_{T_V(K)}^s(\mathbb{F}_p, \mathbb{F}_p)$$

is an isomorphism in degree less or equal to $\ell - s - 1$. Let $T_V(K)_\omega$ be the component of $T_V K$ corresponding to the trivial map. As $Q(K)$ is in $\overline{Nil}_\ell$ the map $K \to (T_V K)_\omega$ is an isomorphism in degrees strictly less to ℓ. This follows from

— $T_V Q(K) = Q(T_V(K)_\omega)$;

— $Q(K) \cong T_V Q(K)$ in degrees $\leq \ell$ since $Q(K)$ is in $\overline{Nil}_\ell$.

As $\mathrm{Tor}_{T_V(K)}^s(\mathbb{F}_p, \mathbb{F}_p) \cong \mathrm{Tor}_{T_V(K)_\omega}^s(\mathbb{F}_p, \mathbb{F}_p)$ it is enough to check that the latter groups are isomorphic to $\mathrm{Tor}_K^s(\mathbb{F}_p, \mathbb{F}_p)$ in degrees less or equal to $\ell - s - 1$. By the above $T_V(K)_\omega$ is isomorphic to $K \oplus J$ where J is an ideal which is $\ell - 1$-connected. The result follows by looking at the standard $T_V(K)_\omega$-resolution of $\mathbb{F}_p$.

Using this lemma one can prove statement (iii) of (8.7.1) exactly as above. Again one can only analyze the kernel K of $E_2^{-1,*} \to E_\infty^{-1,*}$ which turns out to be in $\overline{Nil}_{k+1}$. The cokernel may or may not lie in $\overline{Nil}_{k+1}$. We note without proof the following generalizations of the preceding results. We keep the same assumptions on X.

Proposition 8.7.7. *The kernel of the map*

$$E_2^{-2,*} = \mathrm{Tor}_{H^*X}^{-2}(\mathbb{F}_p, \mathbb{F}_p) \to \Sigma^2 H^*\Omega X / F_{-1}$$

is in $\overline{Nil}_3$.

This is the statement corresponding to (iii) of Theorem 8.7.1 with $k = 0$. The proof is the same as above. However, one cannot go further because one uses (here and above) that there are no differentials starting from $E_2^{-2,*}$ and $E_2^{-1,*}$, which is no longer true for $E_2^{s,*}$ if $s < -2$.

Here is another result:

Theorem 8.7.8 [LS3]. *Let* $F \xrightarrow{i} E \xrightarrow{p} B$ *be a fibration. Assume that all spaces have cohomology of finite type and that* $\pi_1(B)$ *is trivial. Then the kernel of the edge homomorphism of the Eilenberg-Moore spectral sequence* $E_2^{0,*} \cong \mathbb{F}_p \otimes_{H^*B} H^*E \to H^*F$ *is a nilpotent unstable* $\mathcal{A}$*-module.*

The proof is left to the reader. The Eilenberg-Moore spectral sequence has, in this case, similar properties to the above. We note that the result extends to fibre squares. The case of

$$
\begin{array}{ccc}
F \times \Omega B & \xrightarrow{\text{action}} & F \\
\downarrow{\scriptstyle \text{projection}} & & \downarrow \\
F & \longrightarrow & E
\end{array}
$$

is of particular interest.

Proof of 8.7.3 (see also [LS1]*).* The group $\pi_1 X$ is abelian and we are going to show that one can assume it is a p-torsion group. If one is reduced to this case it is enough, by induction, to analyze a covering $\widetilde{X} \to X$ with structural group $\mathbb{Z}/p$ or $\mathbb{Z}/p^\infty$. Obviously one can assume X to be connected.

Consider, if $p = 2$, the case $\mathbb{Z}/2$. We have a fibre square of H-spaces

$$
\begin{array}{ccc}
\widetilde{X} & \longrightarrow & X \\
\downarrow & & \downarrow \\
* & \longrightarrow & B\mathbb{Z}/2 \; .
\end{array}
$$

The Eilenberg-Moore spectral sequence has $E_2^{s,*}$-term

$$\mathrm{Tor}^s_{H^*\mathbb{Z}/2}(H^*X, \mathbb{F}_2) = (H^*X \otimes_{H^*\mathbb{Z}/2} \mathbb{F}_2) \otimes \mathrm{Tor}^s_K(\mathbb{F}_2, \mathbb{F}_2)$$

([MS]). In the above formula K denotes the kernel as Hopf algebra of $H^*\mathbb{Z}/2 \to H^*X$. It is isomorphic to $\mathbb{F}_2[u^{2^s}]$ for some $s \geq 0$. Indeed H^*X being in $\overline{\mathcal{N}il}_k, k \geq 0$, the image of $H^*\mathbb{Z}/2$ in H^*X is finite, therefore, $\mathrm{Tor}^*_K(\mathbb{F}_2, \mathbb{F}_2) = E(v)$ where $\|v\| = (-1, 2^s)$. The action of $\mathcal{A}$ is trivial. It follows that we have an exact sequence of $\mathcal{A}$-modules.

$$0 \to H^*X \otimes_{H^*\mathbb{Z}/2} \mathbb{F}_2 \to H^*\widetilde{X} \to \Sigma^{2^s-1} H^*X \otimes_{H^*\mathbb{Z}/2} \mathbb{F}_2 \to 0$$

(this is the Gysin sequence). Since the canonical map

$$H^*X \to H^*X \otimes_{H^*\mathbb{Z}/2} \mathbb{F}_2$$

has kernel in $\widetilde{\mathcal{N}il}_{k+1}$. The result follows in this case.

As $H^*\mathbb{Z}/2^\infty \cong \mathbb{F}_2[u^2]$, the case of $\mathbb{Z}/2^\infty$ is treated in the same way.

If $p > 2$, and the group is $\mathbb{Z}/p$ consider the classifying map $X \to B\mathbb{Z}/p$ of the covering $\widetilde{X}$ and the fibre $\hat{X}$ of the composite $\varphi : X \to B\mathbb{Z}/p \to BS^1$. As the map $(\times p) \circ \varphi$ is trivial there is a commutative diagram:

$$\begin{array}{ccc} X & \longrightarrow & BS^1 \\ \downarrow & & \downarrow{\times p} \\ * & \longrightarrow & BS^1 . \end{array}$$

Hence we get a map of H-spaces from $\hat{X}$ to S^1. The fibre is easily identified, up to homotopy, with $\widetilde{X}$.

We may assume X connected, thus, $\pi_1 \hat{X} \to \pi_1 S^1$ is surjective. Choose a lifting $S^1 \to X$ of a generator of $\pi_1 S^1$, the composite map $S^1 \times \widetilde{X} \to \hat{X} \times \hat{X} \to \hat{X}$ is a homotopy equivalence. Therefore, if we prove that $H^*\hat{X}$ is in $\widetilde{\mathcal{N}il}_{k+1}$ so will be $H^*\widetilde{X}$. Thus we are left with a fibration $\widetilde{X} \to X \to K(\mathbb{Z}, 2)$ which is analyzed exactly as when $p = 2$. Note, moreover, that this situation is equivalent (in mod p cohomology) to the case of the group $\mathbb{Z}/p^\infty$ and we are done.

We have to explain why one can reduce to the case where $\pi_1 X$ is a p-torsion group. This is essentially classical. First, one can localize at the prime p. From this point, generalities about abelian groups show there are exact sequences:

$$0 \to \mathrm{Tor} \to \pi_1 X \to F \to 0$$

and

$$0 \to \mathbb{Z}_p^{\oplus a} \to F \to \mathbb{Q}^{\oplus \alpha} \to 0 ,$$

where Tor is the torsion subgroup, $a < +\infty$ (because H^*X is of finite type). We can restrict attention to the case $\alpha = 0$. Indeed, the fibre of the canonical map $X \to K(\mathbb{Q}^{\oplus \alpha}, 1)$ as the same mod p cohomology as X. Using that X is an H-space and the same argument as above one observes that X is homotopically equivalent to $X' \times K(\mathbb{Z}^{\oplus a}_{(p)}, 1)$, where X' is the fibre of the canonical map $X \to K(\mathbb{Z}^{\oplus a}, 1)$. Hence if we prove the result for X' it will do for X. We are now done because the fundamental group of X' is a p-torsion group.

8.8. A topological characterization of spaces X such that $H^*X \in \overline{\mathcal{N}il}_k$

Let X be a 1-connected space such that H^*X is of finite type.

Theorem 8.8. *The following two conditions are equivalent:*

(i) H^*X *is in* $\overline{\mathcal{N}il}_k, k \geq 0;$

(ii) *the space of pointed maps* $\mathrm{map}_*(BV, X)$ *is $k - 1$-connected for any V.*

We are only going to prove that (ii) $\Rightarrow$ (i) the other implication being proved as in (8.6). The implication (ii) $\Rightarrow$ (i) could be proved along the same lines as (8.6) but we give a proof depending on (8.7).

We are going to prove that, if $H^*X \in \overline{\mathcal{N}il}_k$ but $H^*X \notin \overline{\mathcal{N}il}_{k+1}$, then $\pi_k \mathrm{map}_*(BV, X)$ is non trivial for some V. Consider the case $k = 0$. In this case, clearly, the result follows from Theorem 8.1.1. Consider now the case $k \geq 1$; $\pi_k \mathrm{map}_*(BV, X)$ is isomorphic to $\left[BV, \Omega^k X\right]_*$. Although $\pi_1 \Omega^{k+1} X$ need not be finite it is easy to check that one can apply Theorem 8.1.1 because $\Omega^k X$ is an H-space. Let $\Omega^k_\bullet X$ be the connected component of the base point of $\Omega^k X$. We conclude that the canonical map $\Omega^k H^*X \to H^* \Omega^k_\bullet X$ has kernel in $\overline{\mathcal{N}il}_1$ (use Theorem 8.7.1 and Proposition 8.7.3). If H^*X is not in $\overline{\mathcal{N}il}_{k+1}$ it means that $\Omega^k H^*X$ is not in $\overline{\mathcal{N}il}_1$. Therefore, $H^* \Omega^k_\bullet X$ is not in $\overline{\mathcal{N}il}_1$ and there is a non-trivial unstable algebra map $H^* \Omega^k_\bullet X \to H^*V$ for some V and we are done by (8.1.1).

9. The generalized Sullivan conjecture and the cohomology of mapping spaces

9.1. The Sullivan conjecture

Let X be a *finite* CW-complex equipped with an action of the group $\mathbb{Z}/p$. We assume that the action is 'nice', for example that X is a $\mathbb{Z}/p$ CW-complex. As usual, we are going to work with the singular complex of X, and get back a theorem about X.

Let $\mathrm{map}_{\mathbb{Z}/p}(E\mathbb{Z}/p, X)$ be the space of $\mathbb{Z}/p$-equivariant maps from the free contractible $\mathbb{Z}/p$-space $E\mathbb{Z}/p$ to X. This space is called the space of **homotopy fixed points** of the action and it is denoted by $X^{h\mathbb{Z}/p}$. An honest fixed point yields an equivariant map $E\mathbb{Z}/p \to X$ and we have a map:

$$X^{\mathbb{Z}/p} \to X^{h\mathbb{Z}/p},$$

by adjointness we get an equivariant map

$$E\mathbb{Z}/p \times X^{\mathbb{Z}/p} \to X.$$

$$E\mathbb{Z}/p \times \mathbb{F}_p^\bullet(X^{\mathbb{Z}/p}) \to \mathbb{F}_p^\bullet(E\mathbb{Z}/p \times X^{\mathbb{Z}/p}) \to \mathbb{F}_p^\bullet(X),$$

and an equivariant map

$$E\mathbb{Z}/p \times \mathbb{F}_{p\infty}(X^{\mathbb{Z}/p}) \to \mathbb{F}_{p\infty}(X),$$

and by adjointness again a map

$$\mathbb{F}_{p\infty}(X^{\mathbb{Z}/p}) \to (\mathbb{F}_{p\infty}X)^{h\mathbb{Z}/p}.$$

Theorem 9.1.1 (Carlsson [Cl2], Lannes [L1][L4], Miller [Ml4]).
The map

$$\mathbb{F}_{p\infty}(X^{\mathbb{Z}/p}) \to (\mathbb{F}_{p\infty}X)^{h\mathbb{Z}/p}$$

is a homotopy equivalence.

The case where the action is trivial corresponds to the original theorem of Miller [Ml2]. In that case the theorem asserts that the canonical map $\mathbb{F}_{p\infty}X \to \text{map}(B\mathbb{Z}/p, \mathbb{F}_{p\infty}X)$ is a homotopy equivalence. This is equivalent to saying that the space of pointed maps $\text{map}_*(B\mathbb{Z}/p, \mathbb{F}_{p\infty}X)$ is contractible. But the 'arithmetic square' of Dror, Dwyer and Kan (see the preceding chapter) shows that, if X is nilpotent and connected, the space $\text{map}_*(B\mathbb{Z}/p, \mathbb{F}_{p\infty}X)$ is weakly equivalent to the space $\text{map}_*(B\mathbb{Z}/p, X)$, implying the contractibility of $\text{map}_*(B\mathbb{Z}/p, X)$. Conversely the contractibility of $\text{map}_*(B\mathbb{Z}/p, X)$ implies the result as stated above if the action is trivial.

The action being still trivial, the result still holds if X is only assumed to be finite dimensional; we refer to [Ml2] for the proof (especially Section 10 to see how one removes the conditions on π_1).

It should be noted that the conditions of Carlsson are a bit different (see [Cl2][Cl3]). His proof is very different from the two others (concerning Miller's approach see also [DMN]).

Next let π be a finite p-group (p prime), X be a finite π-CW-complex. Denote by X^π the space of fixed points under π, by $X^{h\pi}$ the space of π-equivariant maps from the free contractible π-space $E\pi$ to X: $\text{map}_\pi(E\pi, X)$. This space is again called the space of homotopy fixed points. Then following Dwyer and Zabrodsky:

Theorem 9.1.2. *The map*

$$\mathbb{F}_{p\infty}(X^\pi) \to (\mathbb{F}_{p\infty}X)^{h\pi}$$

is a homotopy equivalence.

We are going to present Lannes' proof of Theorem 9.1.1 and explain how to deduce Theorem 9.1.2.

9.2. A comparison theorem

Consider a connected space Y such that H^*Y is of finite type. The following comparison theorem is due to Lannes.

Theorem 9.2.1 [L1][L4]. *Suppose there exist a space Z, such that H^*Z is of finite type, and a map $\omega : BV \times Z \to Y$ such that the map $\widetilde{\omega}^* : T_V H^*Y \to H^*Z$ adjoint to $\omega^* : H^*Y \to H^*V \otimes H^*Z$ is an isomorphism. Then the induced map*

$$\mathbb{F}_{p\infty} Z \to \operatorname{map}(BV, \mathbb{F}_{p\infty} Y)$$

is a homotopy equivalence.

The map ω induces a map

$$\mathbb{F}_{p\infty} Z \times BV \cong \mathbb{F}_{p\infty}(Z \times BV) \to \mathbb{F}_{p\infty} Y .$$

Thus, by adjointness, we get a map

$$\mathbb{F}_{p\infty} Z \to \operatorname{map}(BV, \mathbb{F}_{p\infty} Y) .$$

This is the 'induced map' of the theorem. We show first that it induces a bijection on the π_0. One has a commutative diagram induced by the adjoint $\widetilde{\omega}$ of ω:

$$[BV, \mathbb{F}_{p\infty} Y] \xrightarrow{\ \cong\ } \pi^0 \pi_0 \operatorname{map}(BV, \mathbb{F}_p^\bullet Y) \cong \operatorname{Hom}_{\mathcal{K}}(H^*Y, H^*V)$$

$$\pi_0 \mathbb{F}_{p\infty} Z \cong \pi_0 Z \xrightarrow{\hspace{3cm}} \pi^0 \pi_0(\mathbb{F}_p^\bullet Z) \cong \operatorname{Hom}_{\mathcal{K}}(H^*Z, \mathbb{F}_p)$$

The upper map is a bijection by 8.1.1. The right vertical map identifies with the map $\operatorname{Spec}(\widetilde{\omega}^*) : \operatorname{Spec}(H^0 Z) \to \operatorname{Spec}(T_V^0 H^*Y)$ which is a bijection since $\widetilde{\omega}^*$ is an isomorphism. As $\operatorname{Spec}(H^0 Z)$ identifies with $\pi_0 Z$ the result follows.

We now use the mapping lemma of Bousfield and Kan (see **[BK2]** p. 261).

Lemma 9.2.2. *Let $\{X_n\}$, $n \in \mathbb{N}$, be a (pointed) tower of fibrations such the $E_2^{s,s}$-term of the homotopy spectral sequence is trivial for $s \geq 0$. Let $f : \{X_n\} \to \{Y_n\}$ be a map of tower of fibrations which* *induces an isomorphism on the $E_2^{s,t}$-term of the homotopy spectral sequences for all $t - s \geq 0$. Then f induces a homotopy equivalence*

$$\lim_n X_n \xrightarrow{\sim} \lim_n Y_n \, .$$

Proof of 9.2.1.   Let φ be in $\mathrm{Hom}_{\mathcal{K}}(H^*Y, H^*V)$. Let Z_φ denote the connected component of Z corresponding to φ. Apply Lemma 9.2.2 to the towers

$$\left\{ \mathrm{Tot}_s \, \mathbb{F}_p^\bullet Z_\varphi \right\}_{s \in \mathbb{N}} \, ,$$

$$\left\{ \mathrm{Tot}_s \, \mathrm{map}(BV, \mathbb{F}_p^\bullet Y)_\varphi \right\}_{s \in \mathbb{N}} \, .$$

Recall that $\mathrm{map}(BV, \mathbb{F}_p^\bullet Y)_\varphi$ has been defined in Chapter 8.

It follows from Theorem 7.8.2 that

$$\mathrm{Der}_{\mathcal{K}/H^*V}^s(L, \Sigma^t H^*V) \cong \mathrm{Ext}_{\mathcal{K}_a}^s(T_V L, \Sigma^t \mathbb{F}_p^{\ +}) \, .$$

This together with the fact that $\widetilde{\omega}^*$ is an isomorphism show that the hypotheses on the towers are satisfied. Thus we are done.

9.3. The case of the Borel construction

Let X be a *finite* $\mathbb{Z}/p$-space and Y be the Borel construction on X

$$Y = E\mathbb{Z}/p \underset{\mathbb{Z}/p}{\times} X \, .$$

Recall that this is the quotient of the space $E\mathbb{Z}/p \times X$ by the relations $(eg, x) \sim (e, gx)$. We are going to compute $H^* \mathrm{map}(B\mathbb{Z}/p, \mathbb{F}_{p\infty} Y)$ by using Theorem 9.2.1.

First, observe that the hypothesis on X and the Serre spectral sequence imply that H^*Y is of finite type. Moreover Smith theory implies that $H^* X^{\mathbb{Z}/p}$ is finite dimensional (recall that Cech cohomology and ordinary cohomology agree on X and $X^{\mathbb{Z}/p}$ because they are finite CW-complexes).

Consider now the following space Z. It is the disjoint union over $\varphi \in \mathrm{End}\,\mathbb{Z}/p$ of the spaces Z_φ defined by

— $Z_0 = Y$,
— $Z_\varphi = B\mathbb{Z}/p \times X^{\mathbb{Z}/p}$ if φ is non-trivial.

In order to apply (9.2.1), we need a map $\omega : B\mathbb{Z}/p \times Z \to Y$. This is the disjoint union of the maps $\omega_\varphi : B\mathbb{Z}/p \times Z_\varphi \to Y$ such that:

— ω_0 is the projection on Y,

— if φ is non-zero ω_φ is the composite:

$$B\mathbb{Z}/p \times B\mathbb{Z}/p \times X^{\mathbb{Z}/p} \xrightarrow{\cong} B(\mathbb{Z}/p^{\oplus 2}) \times X^{\mathbb{Z}/p} \to \ldots$$

$$\ldots \xrightarrow{B(\varphi \oplus \mathrm{Id}) \times \mathrm{Id}} B(\mathbb{Z}/p^{\oplus 2}) \times X^{\mathbb{Z}/p} \xrightarrow{B(\mathrm{add}) \times \mathrm{Id}} B\mathbb{Z}/p \times X^{\mathbb{Z}/p} \hookrightarrow Y.$$

We have to check that the induced map $\widetilde{\omega}^* : TH^*Y \to H^*Z$ is an isomorphism.

The group $\mathbb{Z}/p$ acts freely on $X - X^{\mathbb{Z}/p}$. It follows that the space $Y - (B\mathbb{Z}/p \times X^{\mathbb{Z}/p})$ is homotopically equivalent to the finite dimensional space $(X - X^{\mathbb{Z}/p})/\mathbb{Z}/p$. Therefore, the kernel and cokernel of the map

$$H^*\mathbb{Z}/p \otimes H^*X^{\mathbb{Z}/p} \leftarrow H^*Y$$

are finite.

Recall the direct sum decomposition $TM \cong \overline{T}M \oplus M$ (Section 3.2). Proposition 3.3.6 (see also 6.2.1) implies that we have an isomorphism $\overline{T}(H^*\mathbb{Z}/p \otimes H^*X^{\mathbb{Z}/p}) \cong \overline{T}H^*Y$. Again as $H^*X^{\mathbb{Z}/p}$ is finite we have $T(H^*X^{\mathbb{Z}/p}) \cong H^*X^{\mathbb{Z}/p}$. The commutation of T with tensor products yields

$$\overline{T}(H^*\mathbb{Z}/p \otimes H^*X^{\mathbb{Z}/p}) \cong \overline{T}H^*\mathbb{Z}/p \otimes H^*X^{\mathbb{Z}/p}.$$

Therefore, we get an isomorphism

$$\overline{T}H^*(B\mathbb{Z}/p \times X^{\mathbb{Z}/p}) \cong \overline{T}H^*Y.$$

Let us describe the map induced by $\widetilde{\omega}_*$ from $\overline{T}H^*Y$ to $\underset{\varphi \neq 0}{\oplus} H^*Z_\varphi$. Consider the map

$$H^*Y \to H^*Z_\varphi = H^*\mathbb{Z}/p \otimes H^*X^{\mathbb{Z}/p} \xrightarrow{\Phi^* \otimes \mathrm{Id}} H^*(\mathbb{Z}/p^{\oplus 2}) \otimes H^*X^{\mathbb{Z}/p}$$

where $\Phi = B(\mathrm{add} \circ (\varphi \oplus \mathrm{Id}))^*$. The restriction of $\widetilde{\omega}^*$ to $\overline{T}H^*Y$ is the sum of the restrictions to $\overline{T}H^*Y$ of the maps:

$$TH^*Y \xrightarrow{\cong} TH^*Z_\varphi \xrightarrow{\mathrm{adj}(\Phi \times \mathrm{Id})^*} H^*Z_\varphi.$$

$$\bar{T}H^*Y \to \bar{T}H^*(B\mathbb{Z}/p \times X^{\mathbb{Z}/p}) \xrightarrow{\underset{\varphi \neq 0}{\oplus} \operatorname{adj}(\Phi \times \operatorname{Id})^*} \bigoplus_{\varphi \neq 0} H^*Z_\varphi$$

(recall that $Z_\varphi = B\mathbb{Z}/p \times X^{\mathbb{Z}/p}$ for $\varphi \neq 0$).

The right map is an isomorphism. This is shown using the identification $\bar{T}(H^*\mathbb{Z}/p \otimes H^*X^{\mathbb{Z}/p}) \cong (\bar{T}H^*\mathbb{Z}/p) \otimes H^*X^{\mathbb{Z}/p}$ (see above) and the formula $TH^*\mathbb{Z}/p \cong H^*\mathbb{Z}/p^{\operatorname{End}\mathbb{Z}/p}$. Details are left to the reader.

It follows immediately that the map

$$H^*Z \xleftarrow{\widetilde{\omega}^*} TH^*Y$$

is an isomorphism. Therefore

Theorem 9.3. *The map from $\mathbb{F}_{p\infty}Z$ to $\operatorname{map}(B\mathbb{Z}/p, \mathbb{F}_{p\infty}Y)$ adjoint to $\mathbb{F}_{p\infty}\omega$ is a homotopy equivalence.*

9.4. Proof of the Sullivan conjecture

We come back to the proof of Theorem 9.1.1. The space, we are really interested in, is $\operatorname{map}_{\mathbb{Z}/p}(E\mathbb{Z}/p, \mathbb{F}_{p\infty}X)$. What we have described, up to now, is the space $\operatorname{map}(B\mathbb{Z}/p, \mathbb{F}_{p\infty}Y)$, where Y is the Borel construction on X.

The space Y is the total space of a fibration with base $B\mathbb{Z}/p$ and fibre X. It follows from the mod-$\mathbb{F}_p$ fibre lemma (see **[BK2]** p.62) that $\mathbb{F}_{p\infty}Y$ is the total space of a fibration with base $\mathbb{F}_{p\infty}B\mathbb{Z}/p \cong B\mathbb{Z}/p$ and fibre $\mathbb{F}_{p\infty}X$. Note that we can apply the mod-$\mathbb{F}_p$ fibre lemma because $\pi_1 B\mathbb{Z}/p$ acts nilpotently on $H_i X$, these latter groups being by hypothesis finite dimensional $\mathbb{F}_p$-vector spaces (**[BK2]** p.62). In fact $\mathbb{F}_{p\infty}Y$ is homotopically equivalent to the Borel construction on $\mathbb{F}_{p\infty}X$ (note that $\mathbb{F}_{p\infty}X$ has a natural $\mathbb{Z}/p$-action). Thus we have described the space $\operatorname{map}(B\mathbb{Z}/p, E\mathbb{Z}/p \underset{\mathbb{Z}/p}{\times} \mathbb{F}_{p\infty}X)$.

Denote by $\pi : E\mathbb{Z}/p \underset{\mathbb{Z}/p}{\times} \mathbb{F}_{p\infty} X \to B\mathbb{Z}/p$ the projection. The space $\mathrm{map}(B\mathbb{Z}/p, E\mathbb{Z}/p \underset{\mathbb{Z}/p}{\times} \mathbb{F}_{p\infty} X)$ splits up as a disjoint union

$$\coprod_{\varphi \in \mathrm{End}\, \mathbb{Z}/p} \mathrm{map}_\varphi(B\mathbb{Z}/p, E\mathbb{Z}/p \underset{\mathbb{Z}/p}{\times} \mathbb{F}_{p\infty} X).$$

A map f belongs to $\mathrm{map}_\varphi(B\mathbb{Z}/p, E\mathbb{Z}/p \underset{\mathbb{Z}/p}{\times} \mathbb{F}_{p\infty} X)$ if and only if $\Omega(\pi \circ f) = \varphi$ (as an element of $\mathrm{End}\, \mathbb{Z}/p$).

Lemma 9.4.1. *The space* $\mathrm{map}_\varphi(B\mathbb{Z}/p, E\mathbb{Z}/p \underset{\mathbb{Z}/p}{\times} \mathbb{F}_{p\infty} X)$ *is equivalent to the Borel construction on the space* $\mathrm{map}_{\mathbb{Z}/p}(E\mathbb{Z}/p, \mathbb{F}_{p\infty} X)$.

We shall prove a more precise statement and the Sullivan conjecture will follow.

Let W be a $\mathbb{Z}/p$ -space provided with an equivariant map π to $B\mathbb{Z}/p$. Denote by $\mathrm{map}^\varphi_{\mathbb{Z}/p}(E\mathbb{Z}/p, W)$ ($\varphi \in \mathrm{End}\, \mathbb{Z}/p$) the space of $\mathbb{Z}/p$ -equivariant maps $f : E\mathbb{Z}/p \to W$ such that the induced self-map of $B\mathbb{Z}/p$, which is the obvious quotient of $\pi \circ f$, is equal to φ on π_1. One checks that:

$$\mathrm{map}_\varphi(B\mathbb{Z}/p, E\mathbb{Z}/p \underset{\mathbb{Z}/p}{\times} \mathbb{F}_{p\infty} X)$$
$$= \mathrm{map}^\varphi_{\mathbb{Z}/p}(E\mathbb{Z}/p, E\mathbb{Z}/p) \underset{\mathbb{Z}/p}{\times} \mathrm{map}^\varphi_{\mathbb{Z}/p}(E\mathbb{Z}/p, \mathbb{F}_{p\infty} X).$$

It remains to observe that $\mathrm{map}^\varphi_{\mathbb{Z}/p}(E\mathbb{Z}/p, E\mathbb{Z}/p)$ is a model for $E\mathbb{Z}/p$, i.e. a free contractible $\mathbb{Z}/p$ space.

Then one observes that $\widetilde{\omega}$ sends $\mathbb{F}_{p\infty} Z_\varphi \cong B\mathbb{Z}/p \times \mathbb{F}_{p\infty} X^{\mathbb{Z}/p}$ to $\mathrm{map}^\varphi_{\mathbb{Z}/p}(E\mathbb{Z}/p, \mathbb{F}_{p\infty} Y)$. Thus the following diagram is commutative up to homotopy

$$
\begin{array}{ccc}
\mathbb{F}_{p\infty} Z_\varphi & \longrightarrow & \mathrm{map}^\varphi_{\mathbb{Z}/p}(E\mathbb{Z}/p, \mathbb{F}_{p\infty} Y) \\
\downarrow & & \downarrow \\
B\mathbb{Z}/p & \longrightarrow & B\mathbb{Z}/p
\end{array}
$$

Therefore the fibres are equivalent. For $\varphi = \mathrm{Id}$ this gives

$$\mathbb{F}_{p\infty} X^{\mathbb{Z}/p} \cong \mathrm{map}_{\mathbb{Z}/p}(E\mathbb{Z}/p, \mathbb{F}_{p\infty} X).$$

9.5. The case of the action of a finite p-group π

We now explain how to extend Theorem 9.1.1 to any finite p-group
π. The idea is to make an induction on the order of π. As π is
nilpotent it has a non trivial central subgroup isomorphic to $\mathbb{Z}/p$, let
us denote it by σ. We have a short exact sequence:

$$0 \to \sigma \to \pi \to \pi' \to 0 .$$

Let X be a finite π-CW complex. The space X^σ is a finite π'-CW
complex. By Theorem 9.1.1 we know that:

$$\mathbb{F}_{p\infty}(X^\sigma) \cong (\mathbb{F}_{p\infty} X)^{h\sigma} .$$

By the induction hypothesis we can apply Theorem 9.1.2 to the finite
π'-CW complex X^σ. We have:

$$\mathbb{F}_{p\infty}(X^\pi) = \mathbb{F}_{p\infty}\left((X^\sigma)^{\pi'}\right) \cong \left(\mathbb{F}_{p\infty} X^\sigma\right)^{h\pi'} .$$

As $\mathbb{F}_{p\infty} X^\sigma \cong (\mathbb{F}_{p\infty} X)^{h\sigma}$, we get $\left(\mathbb{F}_{p\infty}(X^\sigma)\right)^{h\pi'} \cong \left((\mathbb{F}_{p\infty} X)^{h\sigma}\right)^{h\pi'}$.
The equivalence $\mathbb{F}_{p\infty} X^\sigma \to (\mathbb{F}_{p\infty} X)^{h\sigma}$ is π'-equivariant because
all constructions are functorial and it induces an equivalence on π'-
homotopy fixed point sets by the π'-Whitehead theorem (Proposition 2.7 of [A2]).

It remains to identify, up to homotopy, the space $\left((\mathbb{F}_{p\infty} X)^{h\sigma}\right)^{h\pi'}$
with the space $(\mathbb{F}_{p\infty} X)^{h\pi}$. Take $E\pi$ as a model for $E\sigma$ and note
that $E\pi' \times E\pi$ is also a model for $E\pi$, with π acting diagonally
and through the quotient map $\pi \to \pi'$ on $E\pi'$. We have

$$\mathrm{map}(E\pi' \times E\pi, Z) \cong \mathrm{map}(E\pi', \mathrm{map}(E\pi, Z)) .$$

Passing to π-fixed points by first passing to σ-fixed points and then
to π'-fixed points, we get

$$\mathrm{map}(E\pi' \times E\pi, Z)^\pi \cong \mathrm{map}(E\pi', \mathrm{map}(E\pi, Z)^\sigma)^{\pi'} ,$$

which is $Z^{h\pi} \cong (Z^{h\sigma})^{h\pi'}$.

9.6. The space $\mathrm{map}(BV, BG)$

The following theorem is due to Dwyer-Zabrodsky [**DZ**] and Lannes [**L2**]. Let G be a finite group or more generally a compact Lie group (for more general hypotheses see [**L2**]). Let $\mathrm{rep}(V, G)$ be the finite set of representations of V in G. For a representation $\rho \in \mathrm{rep}(V, G)$ denote by Z_ρ the centralizer in G of ρ, i.e. the subgroup of G consisting of those $g \in G$ such that $g.\rho(v) = \rho(v).g$ for any $v \in V$. The conjugacy class of Z_ρ depends only on ρ. The group homomorphism $(z, v) \mapsto z\rho(v)$, $Z_\rho \times V \to G$ induces a map

$$BZ_\rho \times BV \to BG .$$

There results a map

$$\omega : \Big(\coprod_{\rho \in \mathrm{Rep}(V,G)} BZ_\rho \Big) \times BV \to BG .$$

The following result was soon stated in Chapter 3.

Theorem. *The adjoint map $\widetilde{\omega}^* : T_V H^* BG \to \bigoplus_\rho H^* BZ_\rho$ is an isomorphism.*

The comparison theorem implies:

Theorem 9.6. *The natural map*

$$\coprod_{\rho \in \mathrm{Rep}(V,G)} \mathbb{F}_{p\infty} BZ_\rho \to \mathrm{map}(BV, \mathbb{F}_{p\infty} BG)$$

is a homotopy equivalence.

This too generalizes: we can replace V by any finite p-group π.

In the case of a finite group G the space $\mathrm{map}(BV, BG)$ can be described directly by simplicial method. This is a classical analysis, see for example [**L4**]. It identifies with the space

$$\coprod_{\rho \in \mathrm{Rep}(V,G)} BZ_\rho .$$

In the case of a finite p-group G it follows easily that the unstable module $T_V H^* BG$ is isomorphic to $\bigoplus_\rho H^* BZ_\rho$. We are not going to prove this theorem. However we shall indicate briefly how to prove a particular case. We show how to compute the space $\text{map}(BV, \mathbb{F}_{p\infty} BU(n))$ or the space $\text{map}(BV, \mathbb{F}_{p\infty} BO(n))$. The computations being essentially the same we shall focus on the second example.

Let $V \cong \mathbb{Z}/2^{\oplus n}$ be a maximal elementary abelian 2-group in $O(n)$. The induced map

$$H^* BO(n) \to H^* V$$

is an isomorphism on the invariants under the the action of the Weyl group W of V, i.e. the quotient of the normalizer of V by V. Choose a basis such that V identifies with the diagonal subgroup, H^*V identifies with $\mathbb{F}_2[u_1, \ldots, u_n]$, whilst W identifies with $\mathfrak{S}_n$ acting by permutation of the variables.

It follows from the properties of T that $TH^*BO(n)$ identifies with the invariants under the action of W in $H^*V^{\text{Hom}(\mathbb{Z}/2, V)}$. We leave to the reader as an exercise to check that these invariants are W-equivariant maps from V to H^*V.

Thus, in the given basis, we are looking at $\mathfrak{S}_n$-equivariant maps from V to H^*V. Denote by $(e_1, \ldots, e_n)$ the basis. An equivariant map is determined by its values on the vectors

$$0, \ e_1, \ e_1 + e_2, \ e_1 + e_2 + e_3, \ \ldots, \ e_1 + \ldots + e_n.$$

But these values can freely chosen among the invariants, in H^*V, under the subgroups

$$\mathfrak{S}_n, \mathfrak{S}_1 \times \mathfrak{S}_{n-1}, \ldots, \mathfrak{S}_p \times \mathfrak{S}_{n-p}, \ldots, \mathfrak{S}_n,$$

hence into

$$H^*BO(n), H^*BO(1) \otimes H^*BO(n-1), \ldots$$
$$\ldots, H^*BO(p) \otimes H^*BO(n-p), \ldots, H^*BO(n).$$

Hence there is an isomorphism

$$TH^*BO(n) \cong \bigoplus_{p=0,\ldots,n} H^*BO(p) \otimes H^*BO(n-p).$$

From this point, one identifies the subgroup $O(p) \times O(n-p)$ as the centralizer of an obvious representation of $\mathbb{Z}/2$. It remains to check the conditions of the comparison theorem, essentially to identify the above isomorphism as being the adjoint of the maps induced in cohomology by the maps $B\mathbb{Z}/2 \times BO(p) \times BO(n-p) \to BO(n)$. This is left as an exercise to the reader.

A similar computation could be made for the action of any Coxeter group acting on the symmetric algebra generated by the mod p reduction of the lattice of roots.

9.7. The cohomology of mapping spaces with source BV

The object of this section is to state a theorem due to Lannes [L4] which computes the cohomology of mapping spaces from the classifying space of an elementary abelian p-group. We need some preliminaries in order to state the theorem.

We first describe a natural map

$$T_V H^* X \to H^* \mathrm{map}(BV, \mathbb{F}_{p\infty} X).$$

The evaluation map

$$BV \times \mathrm{map}(BV, X) \to X$$

induces a map

$$H^* X \to H^* V \otimes H^* \mathrm{map}(BV, X),$$

by adjunction we get a map

$$T_V H^* X \to H^* \mathrm{map}(BV, X).$$

In particular, replacing X by $\mathbb{F}_{p\infty} X$, we get a map

$$T_V H^* \mathbb{F}_{p\infty} X \to H^* \mathrm{map}(BV, \mathbb{F}_{p\infty} X).$$

Next the canonical map $H^* \mathbb{F}_{p\infty} X \to H^* X$ has a natural left inverse which we shall describe below. The natural map we are looking for is the composite of the map $T_V H^* X \to T_V H^* \mathbb{F}_{p\infty} X$ this left inverse with the map that we have just described. Let us describe the left inverse.

If $Y^\bullet$ is a cosimplicial space by the very definition of $\mathrm{Tot} Y^\bullet$ there is an evaluation map

$$(\mathrm{Tot} Y^\bullet) \times \Delta^\bullet \to Y^\bullet.$$

This map induces in cohomology a map

$$\pi_0 H^* Y^\bullet \to H^* \mathrm{Tot}\, Y^\bullet .$$

Suppose that $Y^\bullet = \mathbb{F}_p^\bullet X$. Then $\pi_0 H^* \mathbb{F}_p^\bullet X \cong H^* X$, and it is easily checked that the resulting map $H^* X \to H^* \mathbb{F}_{p\infty} X$ is a left inverse for $H^* \mathbb{F}_{p\infty} X \to H^* X$.

Here is the theorem of Lannes [**L4**].

Theorem 9.7.1. *Let X be a space such that the unstable $\mathcal{A}$-algebra $H^* X$ is of finite type. Assume moreover that $T_V H^* X$ is of finite type and trivial in degree 1. Then the natural map*

$$T_V H^* X \to H^* \mathrm{map}(BV, \mathbb{F}_{p\infty} X)$$

is an isomorphism.

The theorem could as well be stated for a connected component of the function space. Let $\varphi : H^* X \to H^* V$ be a morphism of unstable $\mathcal{A}$-algebras. The map that is adjoint to φ determines a $T_V^0 H^* X$-module structure on $\mathbb{F}_p$. Let $(T_V H^* X)_\varphi$ be the unstable $\mathcal{A}$-algebra $T_V H^* X \otimes_{T_V^0 H^* X} \mathbb{F}_p$. Finally let $\mathrm{map}(BV, \mathbb{F}_{p\infty} X)_\varphi$ be the space of maps inducing φ in cohomology (see Chapter 8).

Theorem 9.7.2 [L4]. *Let X be a space such that $H^* X$ is of finite type. Let $\varphi : H^* X \to H^* V$ be morphism of unstable $\mathcal{A}$-algebras. Assume that the unstable $\mathcal{A}$-algebra $(T_V H^* X)_\varphi$ is of finite type and that $(T_V H^* X)_\varphi$ is trivial in degree 1. Then the natural map*

$$(T_V H^* X)_\varphi \to H^* \mathrm{map}(BV, \mathbb{F}_{p\infty} X)_\varphi$$

is an isomorphism.

The proof of Lannes depends on the results of Bousfield on the convergence of the homology spectral sequence of a cosimplicial space [**Bo3**].

The following proposition allows us to state results for the mapping space $\mathrm{map}(BV, X)$.

Proposition 9.7.3 ([DZ], Proposition 3.1). *Let X be a space such that the unstable $\mathcal{A}$-algebra $H^* X$ is of finite type. Then the canonical map*

$$H^* \mathrm{map}(BV, \mathbb{F}_{p\infty} X) \to H^* \mathrm{map}(BV, X)$$

is an isomorphism as soon as $\pi_1 X$ is a finite p-group.

We reproduce below the proof found in [**DZ**].

Proof. It follows from ([**BK2**] Chapter 5) that the canonical map $\pi_1 X \to \pi_1 \mathbb{F}_{p\infty} X$ is an isomorphism. Then the mod-$\mathbb{F}_p$ fibre lemma ([**BK2**] Chapter 2) implies that the universal cover of $\mathbb{F}_{p\infty} X$ is homotopically equivalent to the p-completion of the universal cover of X. Thus the homotopy fibre F of $X \to \mathbb{F}_{p\infty} X$ has an abelian fundamental group. It is a mod p acyclic nilpotent space ([**BK2**] Chapter 2). Now the space $\mathrm{map}(BV, F)$ is mod p acyclic. Indeed, as it is nilpotent and connected, it is enough to show that all its homotopy groups are uniquely p-divisible ([**BK2**] Chapter 5). This is shown using elementary obstruction theory. The result follows.

E. Dror-Farjoun and of J. Smith in [**DS**] gave a very elegant proof of Theorem 9.7.1 that depends on the Eilenberg-Moore spectral sequence. We indicate their proof. Here is the main technical result of Dror-Farjoun and Smith.

Theorem 9.7.4. *Let X be a space such that $H^* X$ is of finite type. Denote by $P_s X$ the s-th Postnikov section of X. Then $T_V H^* X$ is isomorphic to $\mathrm{colim}_s H^*(P_s \mathrm{Tot}_s \mathbb{F}_p^\bullet X)^{BV}$. Moreover if X is nilpotent $T_V H^* X \cong \mathrm{colim}_s H^*(P_s X)^{BV} \cong \mathrm{colim}_s H^*(P_s \mathbb{F}_{p\infty} X)^{BV}$.*

Thus $T_V H^* X$ should be thought of as a kind of continuous cohomology for the space $\mathrm{map}(BV, \mathbb{F}_{p\infty} X)$.

The approach of Dror-Farjoun and Smith is as follows. They observe that the case when X is an Eilenberg-Mac Lane space is classical. Then, using the Eilenberg-Moore spectral sequence, they prove a result for spaces having finitely many non-trivial homotopy groups.

Following Lannes [**L4**] we shall restrict attention to spaces which have only finitely many connected component and, for any choice of the base point, only finitely many non-trivial homotopy groups which will be supposed to be finite p-groups. Lannes called such spaces

p - π_* -finite. Let $F \to E \to B$ be a fibration of p - π_* -finite spaces. Then Dror-Farjoun and Smith prove

Theorem 9.7.5. *If the natural maps*

$$T_V H^* B \to H^* \mathrm{map}(BV, B) \,,$$
$$T_V H^* E \to H^* \mathrm{map}(BV, E)$$

are isomorphisms so is the natural map

$$T_V H^* F \to H^* \mathrm{map}(BV, F) \,.$$

In particular, let X be a connected nilpotent p-local space which has only finitely many non-trivial homotopy groups which are all finite. Then

$$T_V H^* X \cong H^* \mathrm{map}(BV, X) \,.$$

From this point they use a lemma of Bousfield to get Theorems 9.7.4 and 9.7.2.

Note that the method of Dror-Farjoun and Smith gives another proof of Miller's conjecture.

9.8. A special case

When the target space is an Eilenberg-Mac Lane space Theorem 9.7.1 is classical. It is important because we shall do an induction in the Postnikov tower. Thus we handle this case briefly. A generalized Eilenberg-Mac Lane space is a product of Eilenberg-Mac Lane spaces, a simplicial abelian group is a generalized Eilenberg-Mac Lane space.

Proposition 9.8. *Let K be a generalized Eilenberg-Mac Lane space whose mod p cohomology is finitely generated as an $\mathcal{A}$-algebra. In particular this implies that the mod p cohomology is of finite type. Then the natural map*

$$T_V H^* K \to H^* \mathrm{map}(BV, K)$$

is an isomorphism.

Proof. One shows easily that it is enough to prove the proposition in the following cases: $K = K(\mathbb{Z}/p^k, n)$ and $K = K(\mathbb{Z}, n)$. Although not essential, it is also convenient to reduce to the case where $V = \mathbb{F}_p$, the general case follows by induction.

Observe that the space $\mathrm{map}(B\mathbb{Z}/p, K(G, n)$ is a simplicial abelian group. Thus all the connected components have the same homotopy type, they are even isomorphic as simplicial sets. The group of connected components is isomorphic to $H^n(\mathbb{Z}/p, G)$, the ℓ-th homotopy group of any component is isomorphic to $H^{n-\ell}(\mathbb{Z}/p, G)$ by the very definition of Eilenberg-Mac Lane spaces. The component of the unit is a simplicial abelian group, hence a generalized Eilenberg-Mac Lane space. Thus the space $\mathrm{map}(B\mathbb{Z}/p, K(G, n)$ is homotopically equivalent to the product

$$H^n(\mathbb{Z}/p, G) \times \prod_{\ell \geq 1} K(H^{n-\ell}(\mathbb{Z}/p, G), \ell) .$$

Next consider the evaluation map:

$$\mathrm{map}(B\mathbb{Z}/p, K(G, n)) \times B\mathbb{Z}/p \to K(G, n) ;$$

on the factor

$$K(H^{n-\ell}(\mathbb{Z}/p, G), \ell) \times B\mathbb{Z}/p ,$$

it identifies with the map described below. In the case $G = \mathbb{Z}/2$ it is the map that corresponds to the cohomology class $i_\ell \otimes u^{n-i}$, in the case $G = \mathbb{Z}$ it corresponds to the cohomology class $i_\ell \otimes u^{n-i}$ if $n - i$ is even (with an obvious abuse of notation), to the trivial class otherwise. The case of the groups $\mathbb{Z}/2^k$ as well as the case of an odd prime is left to the reader.

We now have to show that the adjoint of the evaluation map induces an isomorphism in cohomology. The main point is to compute $T H^* K(G, n)$. Consider for example the case of $G = \mathbb{Z}$. The mod p cohomology $H^* K(\mathbb{Z}, n)$ is isomorphic to $U(H(n))$, where $H(n)$ is the unstable $\mathcal{A}$-module freely generated by a class i_n of degree n modulo the relation $\beta i_n = 0$ ($\mathrm{Sq}^1 i_n = 0$ if $p = 2$). Recall that U is the Steenrod-Epstein functor. Thus, for all unstable $\mathcal{A}$-modules M, the $\mathbb{F}_p$-vector space $\mathrm{Hom}_{\mathcal{U}}(H(n), M)$ identifies naturally with the kernel of the Bockstein homomorphism $\beta : M^n \to M^{n+1}$. An easy computation shows that:

$$T(H(2n)) \cong H(2n) \ \oplus \ \bigoplus_{i=0,\ldots,n-1} F(2i),$$

and that

$$T(H(2n+1)) \cong H(2n+1) \ \oplus \ \bigoplus_{i=0,\ldots,n-1} F(2i+1).$$

Using the commutation of the functor T with the functor U this allows us to compute $T(H^*(K(G,n)))$ and to check the announced isomorphisms.

It should be observed that this theorem is false without the hypothesis that H^*K is finitely generated as an $\mathcal{A}$-algebra. Consider for example the case of $K = \prod_{n\geq 0} K(\mathbb{Z}/p, n)$. Obviously the mod p cohomology is not finitely generated as an $\mathcal{A}$-algebra. The unstable $\mathcal{A}$-algebra $T_V H^*K$ is easily checked to be of countable dimension, as an $\mathbb{F}_p$-vector space, in all degrees. But an $\mathbb{F}_p$-vector space of countable dimension cannot be a dual. Thus the theorem cannot hold for this space.

Nevertheless it is easy to check that the map

$$T_V H^*K \to H^* \mathrm{map}(BV, K)$$

is always injective (K being still a generalized Eilenberg-Mac Lane space).

9.9. The Eilenberg-Moore spectral sequence

We prove in this section Theorem 9.7.5 that will allow us to make induction on the Postnikov tower.

Recall that we restrict attention to p-π_*-finite spaces, i.e. spaces which have only finitely many connected component and only finitely many non-trivial homotopy groups which are supposed to be finite p-groups. Let

$$F \to E \to B$$

be a fibration of p-π_*-finite spaces. We recall the statement:

Theorem. *If the natural maps*

$$T_V H^* B \to H^* \mathrm{map}(BV, B),$$
$$T_V H^* E \to H^* \mathrm{map}(BV, E)$$

are isomorphisms so is the natural map

$$T_V H^* F \to H^* \mathrm{map}(BV, F).$$

In particular, let X be a nilpotent p-local space which has only finitely many non-trivial homotopy groups which are all finite. Then

$$T_V H^* X \cong H^* \mathrm{map}(BV, X).$$

Proof. We suppose that the theorem holds for E and B and we prove it for F. As the fundamental group of B is a finite p-group it acts nilpotently on the homology of the fibre and the Eilenberg-Moore spectral sequence of the fibration converges to $H^* F$ **[D]**.

Applying the functor $\mathrm{map}(BV, -)$ to the preceding fibration we get a new fibration

$$\mathrm{map}(BV, F) \to \mathrm{map}(BV, E) \to \mathrm{map}(BV, B).$$

We apply the Eilenberg-Moore spectral sequence to this fibration to compute the cohomology of the fibre over the map that sends BV to the base point of B. This fibre is $\mathrm{map}(BV, F)$. The E_2-term of the spectral sequence is:

$$\mathrm{Tor}^s_{H^* \mathrm{map}(BV, B)}(H^* \mathrm{map}(BV, E), \mathbb{F}_p);$$

which is by hypothesis isomorphic to

$$\mathrm{Tor}^s_{T_V(H^* B)}(T_V(H^* E), \mathbb{F}_p);$$

which is itself isomorphic to

$$T_V(\mathrm{Tor}^s_{H^* B}(H^* E, \mathbb{F}_p)).$$

by Lemma 6.4.2. This spectral sequence converges for the same reason has above. The fundamental group of the base is a finite p-group and the fibre has mod p homology of finite type. This is proved by using elementary homotopy theory.

On the other hand it is possible to apply the functor T_V to the E_2-term of the Eilenberg-Moore spectral sequence of the initial fibration

$F \to E \to B$. Indeed this E_2-term is an unstable $\mathcal{A}$-module and the differential are $\mathcal{A}$-linear (see Section 8.7). Hence, we get a new spectral sequence whose E^2-term is isomorphic to $T_V(\mathrm{Tor}_{H^*B}(H^*E, \mathbb{F}_p))$. The evaluation maps and the definition of the functor T_V determine a natural map from this new spectral sequence to the Eilenberg-Moore spectral sequence of the mapping spaces. This map is an isomorphism on the E_2-term. Consequently it is is an isomorphism on the E_∞-term. Then, using the exactness of T_V, the fact it commutes with colimits and the convergence of the spectral sequences we get that the map

$$T_V H^* F \to H^* \mathrm{map}(BV, F)$$

is an isomorphism.

Proof of theorem 9.7.4.   Let X be a space such that $H^* X$ is of finite type. Consider the space $X_s = P_s \mathrm{Tot}_s \mathbb{F}_p^\bullet X$, it is p-π_*-finite.

From now on we shall use mod p homology. We shall denote $H_*(X; \mathbb{F}_p)$ by $H_* X$.

Recall a few definitions. A tower of groups $\{G_n, f_n\}$, $n \in \mathbb{N}$, is constant if all G_n's are equal and if the maps f_n are equal to the identity for all n. A tower of groups $\{G_n, f_n\}$, $n \in \mathbb{N}$, is said to be pro-trivial if for any n there exists p such that the composite map $f_{n+p} \circ \cdots \circ f_{n+1} : G_{n+p} \to G_n$ is trivial. A map between two towers of groups is a pro-isomorphism if the kernel and the cokernel are pro-trivial. The limits as well as the $\lim^1$ of two pro-isomorphic towers are isomorphic. Finally, a tower is pro-constant if it is pro-isomorphic to a constant tower.

It follows from [**Dr**] that the canonical map from the constant tower $\{H_i X\}$ to the tower $\{H_i \mathrm{Tot}_s \mathbb{F}_p^\bullet X\}$, $s \in \mathbb{N}$, is a pro-isomorphism. Hence (see [**BK2**], Chapter 3, Proposition 3.2, Corollary 3.4), it is also pro-isomorphic to the tower $\{H_i X_s\}$, $s \in \mathbb{N}$. This implies that $H^* X$ is isomorphic to $\mathrm{colim}_s H^* X_s$. As T_V commutes with colimits we have

$$T_V H^* X \cong \mathrm{colim}_s H^* \mathrm{map}(BV, X_s).$$

We leave it the reader to complete the proof of the theorem in the case of a nilpotent space.

9.10. A lemma of Bousfield

Lemma 9.10.1. *Let $\{X_n, g_n\}$, $n \in \mathbb{N}$, be a tower of fibrations of pointed $\mathbb{F}_p$-nilpotent spaces. If the tower of $\mathbb{F}_p$-vector spaces $\{\widetilde{H}_i X_n, g_{n*}\}$, $n \in \mathbb{N}$, is pro-trivial for $i \leq 1$ and pro-constant for $i > 1$ the space $\lim_{n \in \mathbb{N}} X_n$ is simply connected and the map from the constant tower*

$$\{H_i \lim_{n \in \mathbb{N}} X_n\} \to \{H_i X_n, g_{n*}\}$$

is a pro-isomorphism for all i.

We reproduce here the proof of Bousfield (see **[Bo3]**).

Proof. Since the tower $\{\widetilde{H}_i X_n, g_{n*}\}$, $n \in \mathbb{N}$, is pro-trivial for $i \leq 1$, and since the spaces X_n are $\mathbb{F}_p$-nilpotent the tower $\{\pi_i X_n, g_{n*}\}$ is pro-trivial for $i \leq 1$. It follows easily that the tower $\{X_n, g_n\}$ is weakly pro-homotopy equivalent to the tower $\{\widetilde{X}_n, \widetilde{g}_n\}$, $n \in \mathbb{N}$, of universal covers. This means that the corresponding towers of homotopy groups are pro-isomorphic. In particular $\lim_n X_n$ is homotopically equivalent to the limit $\lim_n \widetilde{X}_n$ (**[BK2]** Chapter 3, although we do not assume that the X_n's are connected the result there extends because of the pro-triviality of the tower $\{\widetilde{H}_0 X_n, g_{n*}\}$). Thus we may suppose that each X_n is simply connected. Now, since the tower $\{H_2 X_n\}$ is pro-constant, we have

$$\lim_n{}^1 \pi_2 X_n = 0.$$

Hence (**[BK2]** Chapter 9) $\lim_n X_n$ is simply connected.

Let F_n be the fibre of the map

$$\lim_n X_n \to X_n.$$

Each F_n is simple and has an abelian fundamental group.

Using the Eilenberg-Moore spectral sequence it is easily shown that, for any i, the tower $\{H_i F_n\}$, $n \in \mathbb{N}$, is pro-constant. As $\lim_n F_n$ is a point we get (**[BK2]** Chapter 9)

$$\lim_n \pi_i F_n = \lim_n{}^1 \pi_i F_n = 0,$$

and thus

$$\lim_n \pi_i F_n \otimes \mathbb{Z}/p = \lim_n \mathrm{Tor}(\pi_i F_n, \mathbb{Z}/p) = 0, \text{ for all } i.$$

Consequently, since the towers $\{\widetilde{H}_1 F_n\}$, and $\{\widetilde{H}_2 F_n\}$ are pro-constant, the towers $\{\pi_1 F_n \otimes \mathbb{Z}/p\}$, and $\{\mathrm{Tor}(\pi_1 F_n, \mathbb{Z}/p)\}$ are pro-trivial. Hence ([**BK2**], Chapter 3, Corollary 3.4), the towers $\{\widetilde{H}_i K(\pi_1 F_n, 1)\}$ are pro-trivial for all i. If $\widetilde{F}_n$ denotes the universal cover of F_n, the towers $\{\widetilde{H}_i \widetilde{F}_n\}$, and $\{\widetilde{H}_i F_n\}$ are pro-isomorphic by a Serre spectral sequence argument. Proceeding inductively we deduce that $\{\widetilde{H}_i F_n\}$ is pro-trivial, and that $\{\widetilde{H}_i \lim_n X_n\}$ is pro-isomorphic to $\{\widetilde{H}_i X_n\}$ for all i as required.

Proof of theorem 9.7.2.

It remains to apply Bousfield's lemma. The tower $\{X_s\}$, $s \in \mathbb{N}$ (recall that $X_s = P_s \mathrm{Tot}_s \mathbb{F}_p^\bullet X$) is a tower of $\mathbb{F}_p$-nilpotent spaces. We have to check that the tower

$$\{\widetilde{H}_i(\mathrm{map}(BV, X_s); \mathbb{F}_p)_\varphi\}_{s \in \mathbb{N}}$$

is pro-trivial if $i \leq 1$, pro-constant if $i > 1$ (note that the case $i = 0$ is obvious).

It is a consequence from the following easy fact. Let $\{H_s\}$, $s \in \mathbb{N}$, be a tower of finite groups $\{H_s\}$ that has a finite limit. Then the map from the constant tower $\{\lim_s H_s\}$ to the tower $\{H_s\}$ is a pro-isomorphism. The case $i = 1$ follows from the hypothesis that $T_V^1 H^*(X_s)_\varphi = 0$. As taking limits commutes with taking function spaces we get finally

$$T_V(H^* X)_\varphi \cong H^* \mathrm{map}(BV, \mathbb{F}_{p\infty} X)_\varphi.$$

Again, if X is nilpotent, $\mathbb{F}_{p\infty} X$ can be replaced by X.

References

[A1] J.F. ADAMS, *Two Theorems of J. Lannes*, in Collected Works, Editors J.P. May, C.B. Thomas, Camb. Univ. Press 1992.

[A2] J.F. ADAMS, *Prerequisites (on equivariant theory) for Carlsson's lecture*, Algebraic Topology, Proc. Aarhus Conf. 1982, Springer Lect. Notes in Math. 1051 (1984), 483-532.

[AGM] J.F. ADAMS, J.H. GUNAWARDENA, H.R. MILLER, *The Segal conjecture for elementary abelian p -groups*, Topology 24 (1985), 435-460.

[An] M. ANDRE, *Homologie des algèbres commutatives*, Grundlehren der Mathematischen Wissenchaften 206, Springer-Verlag, 1974.

[Aw] G. ANDREWS, *The Rogers-Ramanujan reciprocal and Minc's partition function*, Pacific J. of Math. 95 (1981), 251-256.

[AW] J.F. ADAMS, C.W. WILKERSON, *Finite H-spaces and algebras over the Steenrod algebra*, Ann. of Math. 111 (1980), 95-143.

[BB] M. BARR, J. BECK, *Homology and standard constructions*, in Seminar on triples and categorical homology theory, Springer Lecture Notes in Mathematics 80 (1969), 245-335.

[BC] E.H BROWN, R.L. COHEN, *The Adams spectral sequence of $\Omega^2 S^3$ and Brown-Gitler spectra, Algebraic Topology and Algebraic K -theory*, Annals of Math. Studies 113, Princeton University Press (1987), 101-125.

[Be] D. BENSON, *The Loewy structure of the projective indecomposable modules for A_8 in characteristic 2*, Comm. in Alg. 11 (13) (1983), 1395-1432.

[BF] D. BENSON, V. FRANJOU, *Séries de compositions de modules instables et injectivité de la cohomologie du groupe $\mathbb{Z}/2$* , Math. Z. 208 (1991), 389-399.

[BG] E.H. BROWN, S. GITLER, *A spectrum whose cohomology is a certain cyclic module over the Steenrod algebra*, Topology 12 (1973), 283-295.

[BK1]   A.K. BOUSFIELD, D.M. KAN, *The homotopy spectral sequence of a space with coefficients in a ring*, Topology 11 (1972), 76-106.

[BK2]   A.K. BOUSFIELD, D.M. KAN, *Homotopy limits, completions and localizations*, Springer Lecture Notes in Math. 304, 1972.

[Bo1]   A.K. BOUSFIELD, *Nice homology coalgebras*, Trans. Amer. Math. Soc. 148 (1970), 473-489.

[Bo2]   A.K. BOUSFIELD, *The localization of spaces with respect to homology*, Topology 14 (1975), 133-150.

[Bo3]   A.K. BOUSFIELD, *On the homology spectral sequence of a cosimplicial space*, Amer. J. of Math. 109 (1987), 361-394.

[Bo4]   A.K. BOUSFIELD, *Homotopy spectral sequences and obstructions*, Israël J. of Math. 66 (1989), 54-104.

[Br]   N. BOURBAKI, Algèbre, Chapitre 10, Paris, Hermann, 1980.

[BZ1]   C. BROTO, S. ZARATI, *Nil-localization of unstable algebras over the Steenrod algebra*, Math. Z. 199 (1988), 525-537.

[BZ2]   C. BROTO, S. ZARATI, *On sub- $\mathcal{A}_p^*$ algebra of H^*V* , Proc. Barcelona 1990, Springer Lecture Notes in Mathematics 1509, 35-49.

[C]   H. CARTAN, *Sur les groupes d'Eilenberg-Mac Lane II*, Proc. Nat. Acad. Sci. USA 40 (1954), 704-707.

[CE]   H. CARTAN, S. EILENBERG, *Homological algebra*, Princeton Univ. Press, 1956.

[Ch]   R.L. COHEN, *Odd primary infinite families in stable homotopy*, Mem. Amer. Math. Soc. 242 (1981).

[CK1]   D. CARLISLE, N. KUHN, *Subalgebras of the Steenrod algebra and the action of matrices on truncated polynomial algebras*, J. Algebra 121 (1989), 370-387.

[CK2]   D. CARLISLE, N. KUHN, *Smah products of summands of $B(\mathbb{Z}/p)^n$* , A.M.S Cont. Math. 96, 87-102.

[Cl1]   G. CARLSSON, *G.B. Segal's Burnside ring conjecture for $(\mathbb{Z}/2)^k$* , Topology 22 (1983), 83-103.

[Cl2] G. CARLSSON, *Segal's Burnside ring conjecture and the homotopy limit problem*, Proc. Durham Symp. on Homotopy Theorey 1985, L.M.S. Camb. Univ. Press (1987), 6-34.

[Cl3] G. CARLSSON, *Equivariant stable homotopy and Sullivan's conjecture*, Invent. Math. 103 (1991), 497-525.

[Cl4] G. CARLSSON, *Equivariant stable homotopy and Segal's Burnside ring conjecture*, Annals of Math. 120 (1984), 189-224.

[CR] C. CURTIS, I. REINER, *Representation theory of finite groups and associative algebra*, J. Wiley 1962.

[CS] H.E.A. CAMPBELL, S.P. SELICK, *Polynomial algebras over the Steenrod algebra*, Comment. Math. Helv. 65 (1990), 171-180.

[CW] D. CARLISLE, G. WALKER, *Poincaré Series for the occurence of certain modular representations of $GL(n, p)$ in the symmetric algebra*, Proc. of the Royal Society of Edinburgh 113A (1989), 27-41.

[Da] D. DAVIS, *A family of unstable Steenrod-modules which includes those of G. Carlsson*, Journal of Pure and Appl. Alg. 35 (1985), 253-267.

[D] W.G. DWYER, *Strong convergence of the Eilenberg-Moore spectral sequence*, Topology 13 (1974), 255-265.

[DDK] E. DROR, W.G. DWYER, D.M. KAN, *An arithmetic square for virtually nilpotent spaces*, Ill. J. of Math. 21 (1977), 242-254.

[DLS] J. DUFLOT, P.S. LANDWEBER, R.E. STONG, *A problem on $H^*(BG; \mathbb{Z}/p)$*. Algebraic Topology Proc. 1984, Gottingen, Springer Lect. Notes in Math 1172 (1985).

[DMN] W.G. DWYER, H.R. MILLER, J.A. NEISENDORFER, *Fibrewise completion and unstable Adams spectral sequence*, Israël J. of Math. 66 (1989), 160-178.

[DMW] W.G. DWYER, H.R. MILLER, C. WILKERSON, *The homotopy uniqueness of BS^3*, Algebraic Topology Proc. 1986, Barcelona, Springer Lect. Notes in Math 1298 (1987), 90-105.

[Do] A. DOLD, *Homology of symmetric products and other functors of complexes*, Annals of Math. 68 (1958), 54-80.

[Dr] E. DROR FARJOUN, *Pro-nilpotent representation of homology types*, Proc. Am. Soc. 38 (1973), 657-660.

[DS] E. DROR FARJOUN, J. SMITH, *A geometric interpretation of Lannes' functor* T , Proc. Luminy 1988, Astérisque 191 (1990), 87-95.

[DtW] S. DOTY, and G. WALKER, *The composition factors of* $\mathbb{F}_p[x_1, x_2, x_3]$ *as a* $GL(3, p)$ *-module*, Journal of Algebra 146 (1992).

[DW1] W.G. DWYER, C.W. WILKERSON, *Smith theory and the functor* T , Comm. Math. Helv. 66 (1991), 1-17.

[DW2] W.G. DWYER, C.W. WILKERSON, *Spaces of null homotopic maps*, Proc. Luminy 1988, Astérisque 191 (1990), 97-108.

[DW3] W.G. DWYER, C.W. WILKERSON, *A new finite H-space at the prime* 2 , preprint 1989.

[DW4] W.G. DWYER, C.W. WILKERSON, *A cohomology decomposition theorem*, Topology 31 (1992), 433-443.

[DZ] W.G. DWYER, A. ZABRODSKY, *Maps between classifying spaces*, Algebraic Topology, Proc. Barcelona 1986, Springer Lect. Notes in Math. 1298 (1987), 106-119.

[EM] S. EILENBERG, S. MAC LANE, *On the groups* $H(\Pi, n)$ *, II*, Annals of Math. 60 (1954), 49-139.

[F] P. FREYD, *Abelian categories : an introduction to the theory of functors*, New-York, Harper and Row, 1964.

[FP] P. FLAJOLET, H. PRODINGER, *Level number sequences for trees*, Discrete Math. 65 (1987), 149-156.

[FS] V. FRANJOU, L. SCHWARTZ, *Reduced unstable* $\mathcal{A}$ *-modules and the modular representation theory of the symmetric groups*, Ann. Scient. Ec. Norm. Sup. 23 (1990), 593-624.

[Gb] P. GABRIEL, *Des catégories abéliennes*, Bull. Soc. Math. France 90 (1962), 323-348.

[Gd] R. GODEMENT, *Théorie des faisceaux*, Hermann, Paris, 1964.

[Ge1] P. GOERSS, *On the structure of the André-Quillen cohomology of commutative* $\mathbb{F}_2$ *-algebras*, Astérisque 186 (1990).

[Ge2]   P. GOERSS, *André-Quillen cohomology and the homotopy groups of mapping spaces : understanding the E_2 -term of the Bousfield-Kan spectral sequence*, J. Pure and Appl. Alg. 63 (1990), 113-153.

[H1]   H.-W. HENN, *Classifying spaces with injective mod p cohomology*, Comment. Math. Helv. 64 (1989), 200-205.

[H2]   H.-W. HENN, *Some finiteness results in the category of unstable modules over the Steenrod algebra and applications to stable splittings*, Math. Ann. 291 (1991), 191-203.

[HLS1] H.-W. HENN, J. LANNES, L. SCHWARTZ, *Analytic functors, unstable algebras and cohomology of classifying spaces*, Algebraic Topology Proc. 1988, Northwestern University, Cont. Math. 96 (1989), 197-220.

[HLS2] H.-W. HENN, J. LANNES, L. SCHWARTZ, *The categories of unstable modules and unstable algebras modulo nilpotent objects*, To appear in Am. J. of Math. 1989.

[HLS3] H.-W. HENN, J. LANNES, L. SCHWARTZ, *Localizations of unstable $\mathcal{A}$ -modules and equivariant mod p -cohomology*, Preprint IHES, 1992.

[HK]   J.C. HARRIS, N.J. KUHN, *Stable decompositions of classifying spaces of finite abelian p -groups*, Math. Proc. Camb. Phil. Soc. 103 (1988), 427-449.

[HS]   H.-W. HENN, L. SCHWARTZ, *Indecomposable $\mathcal{A}$ -module summands in H^*V which are unstable algebras*, Math. Z. 205 (1990), 145-158.

[I]   K. ISHIGURO, *Unstable Adams operations on classifying spaces*, Math. Proc. of the Camb. Phil. Soc.102 (1987), 71-75.

[J]   G. JAMES, *The decomposition of tensors over fields of prime characteristic*, Math. Z. 172 (1980), 161-178.

[JK]   G. JAMES, A. KERBER, *The representation theory of the symmetric groups*, Encycl. Math. Appl. 16, 1981.

[JM]   S. JACKOWSKY, J. McCLURE, *A homotopy decomposition theorem for classifying spaces of compact Lie groups*, Topology 31 (1992), 113-132.

[JMO] S. JACKOWSKY, J. McCLURE, R. OLIVER, *Self-maps of classifying spaces of compact simple Lie groups*, Bull. Am. Math. Soc. 22 (1990), 65-72; and Ann. of Math. 136 (1991).

[K1] N.J. KUHN, *Generic Representations of the Finite General Linear Groups and the Steenrod Algebra I and II*, Preprint, Charlottesville (1990) and (1992), Part I is to appear in Am. J. of Math.

[K2] N.J. KUHN, *Generic Representations of the Finite General Linear Groups and Lannes' T-functor*, L. M. S. Lecture Notes 176.

[KM] N.J. KUHN, S. MITCHELL, *The multiplicity of the Steinberg representation of* $GL_n \mathbb{F}_q$ *in the symmetric algebra*, Proc. Am. Math. Soc. 96 (1986), 1-6.

[L1] J. LANNES, *Sur la cohomologie modulo* p *des* p *-groupes abéliens élémentaires*, Proc. Durham Symp. on Homotopy Theory 1985, LMS, Camb. Univ. Press, 1987, 97-116.

[L2] J. LANNES, *Cohomology of groups and function spaces*, Preprint 1986.

[L3] J. LANNES, *Sur le n dual du n-ème spectre de Brown-Gitler*, Math. Z. 159 (1988), 29-42.

[L4] J. LANNES, *Sur les espaces fonctionnels dont la source est le classifiant d'un* p *-groupe abélien élémentaire*, Publ. I.H.E.S. 75 (1992), 135-244.

[Li] W. H. LI, *Iterated loop functors and the homology of the Steenrod algebra*, Thesis Fordham University, New York 1980.

[LS1] J. LANNES, L. SCHWARTZ, *A propos de conjectures de Serre et Sullivan*, Invent. Math. 83 (1986), 593-603.

[LS2] J. LANNES, L. SCHWARTZ, *Sur la structure des* $\mathcal{A}$ *-modules instables injectifs*, Topology 28 (1989), 153-169.

[LS3] J. LANNES, L. SCHWARTZ, *Sur les groupes d'homotopie des espaces dont la cohomologie modulo 2 est nilpotente*, Israël J. of Math. 66 (1989), 260-273.

[LS4] J. LANNES, L. SCHWARTZ, *Modules instables sur l'algèbre de Steenrod et foncteurs analytiques*, notes manuscrites, (1986).

[LZ1] J. LANNES, S. ZARATI, *Sur les* $\mathcal{U}$ *-injectifs*, Ann. Scient. Ec. Norm. Sup. 19 (1986), 1-31.

[LZ2] J. LANNES, S. ZARATI, *Foncteurs dérivés de la déstabilisation*, Math. Z. 194 (1987), 25-59.

[M] J.P. MAY, *Simplicial objects in algebraic topology*, Van Nostrand Math. Studies 11, 1967; reprinted by the University of Chicago Press, 1992.

[Md] I. G. MACDONALD, *Symmetric Functions and Hall Polynomials*, Oxford Mathematical Monographs, Clarendon Press, 1979.

[Mh] M. MAHOWALD, *A new infinite family in* $_2\pi_* S$, Topology 16 (1977), 249-254.

[ML] S. MAC LANE, *Categories for the working mathematician*, Graduate text in Mathematics 5, Springer, 1971.

[Ml1] H.R. MILLER, *Massey-Peterson towers and maps from classifying spaces*, Algebraic topology Aarhus 1982 (proceedings), Springer Lect. Notes in Math. 1051 (1984), 401-417.

[Ml2] H.R. MILLER, *The Sullivan conjecture on maps from classifying spaces*, Annals of Math. 120 (1984), 39-87; and corrigendum, Annals of Math. 121 (1985), 605-609.

[Ml3] H.R. MILLER, Letter 1985.

[Ml4] H.R. MILLER, *The Sullivan conjecture and homotopical representation theory*, Proc. Internat. Congr. Math. Berkeley 1986, 580-589.

[MM] J. MILNOR, J. MOORE, *On the structure of Hopf algebras*, Ann. of Math. (2) 81 (1965), 211-264.

[Mn] J. MILNOR, *The Steenrod algebra and its dual*, Ann. of Math. 67 (1958), 150-171.

[MP] W.S. MASSEY, F. PETERSON, *The mod* 2 *cohomology structure of certain fibre spaces*, Mem. of the Amer. Math. Soc. 74 (1967).

[Mt1] S. MITCHELL, *Splitting* $B(\mathbb{Z}/p)^n$ *and* $B\mathbb{T}^n$ *via modular representation theory*, Math. Z. 189 (1985), 1-9.

[Mt2]   S. MITCHELL, *Finite complexes with $A(n)$ -free cohomology*,
        Topology 24 (1985), 227-246.

[Mt3]   S. MITCHELL, *On the Steinberg module, representations of the
        symmetric groups and the Steenrod algebra*, J. Pure and Appl.
        Alg. 39 (1986), 275-281.

[MtP]   S. MITCHELL, S. PRIDDY, *Stable splittings derived from the
        Steinberg module*, Topology 22 (1983), 285-298.

[MS]   J. MOORE, L. SMITH, *Hopf algebras and multiplicative fibra-
        tions I*, Am. J. of Math. 90 (1968), 752-780.

[Na]   M. NAKAOKA, *Cohomology mod p of symmetric power of
        spheres*, J. Inst. Poly. Osaka City Univ. 9 (1958), 1-18.

[Ni]   G. NISHIDA, *Stable homotopy type of classifying spaces of finite
        groups*, Preprint (1985).

[Q1]   D. QUILLEN, *On the homology of commutative rings*, Proc.
        Symp. Pure Math. 17 (1970), 65-87.

[Q2]   D. QUILLEN, *The spectrum of an equivariant cohomology ring,
        I, II*, Ann. of Math. 94 (1971), 549-572; 573-602.

[R]   D. RECTOR, *Steenrod operations in the Eilenberg-Moore spec-
        tral sequence*, Comm. Math. Helv. 45 (1970), 540-552.

[S1]   J.-P. SERRE, *Cohomologie modulo 2 des complexes
        d'Eilenberg-Mac-Lane*, Comm. Math. Helv. 27 (1953),
        198-232.

[S2]   J.-P. SERRE, *Représentations linéaires des groupes finis*, Her-
        mann, Paris, 1971.

[Sc1]   L. SCHWARTZ, *La filtration nilpotente de la catégorie U et
        la cohomologie des espaces de lacets*, Proc. Louvain la Neuve
        1986, Springer Lect. Notes in Math. 1318 (1988), 208-218.

[Sc2]   L. SCHWARTZ, *L'anneau de Grothendieck de la categorie U*,
        Notes 1991.

[SE]   N.E. STEENROD, D.B.A. EPSTEIN, *Cohomology operations*,
        Annals of Math. Studies 50, Princeton Univ. Press, 1962.

[Sh]   R. SHANK *Lannes' T functor on primitively generated Hopf
        algebras*, Math. Z. 211 (1992), 341-350.

[Si] W.M. SINGER, *Iterated loop functors and the homology of the Steenrod algebra*, J. Pure and Appl. Alg. 11 (1977), 83-101.

[Sm] L. SMITH, *On Kunneth Theorem I, the Eilenberg-Moore spectral sequence*, Math. Z. 116 (1970), 94-140.

[Sm] L. SMITH, R. SWITZER, *Polynomial algebras over the Steenrod algebra. Variations on a theorem of Adams and Wilkerson*, Proc. Edinburgh Math. Soc. 27 (1984), 11-19.

[Su] D. SULLIVAN, *Genetics of homotopy theory and the Adams conjecture*, Annals of Math. 100 (1974), 1-79.

[T1] M. TANGORA, *Computing the homology of the lambda algebra*, Mem. Amer. Math. Soc. 337 (1985).

[T2] M. TANGORA, *Level number sequences of trees and the lambda algebra*, Preprint 1990.

[W] R. WOOD, *Splitting $\Sigma(\mathbb{C}P^\infty \times ... \times \mathbb{C}P^\infty)$ and the action of Steenrod squares Sq^i on the polynomial ring $\mathbb{F}_2[x_1, ..., x_n]$*, Algebraic Topology Proc. 1986, Barcelona, Springer Lect. Notes in Math 1298 (1987), 237-255.

[Z] S. ZARATI, *Dérivés du foncteur de déstabilisation en caractéristique impaire et applications*, Thèse de Doctorat d' Etat, Orsay 1984.

Index of notation

Index